SMALL ISLAND AND SMALL DESTINATION TOURISM

Overcoming the Smallness Barrier for
Economic Growth and Tourism Competitiveness

Advances in Hospitality and Tourism Series

SMALL ISLAND AND SMALL DESTINATION TOURISM

Overcoming the Smallness Barrier for Economic Growth and Tourism Competitiveness

Robertico Croes, PhD

AAP APPLE
ACADEMIC
PRESS

First edition published 2023

Apple Academic Press Inc.
1265 Goldenrod Circle, NE,
Palm Bay, FL 32905 USA
4164 Lakeshore Road, Burlington,
ON, L7L 1A4 Canada

CRC Press
6000 Broken Sound Parkway NW,
Suite 300, Boca Raton, FL 33487-2742 USA
4 Park Square, Milton Park,
Abingdon, Oxon, OX14 4RN UK

© 2023 by Apple Academic Press, Inc.

Apple Academic Press exclusively co-publishes with CRC Press, an imprint of Taylor & Francis Group, LLC

Library and Archives Canada Cataloguing in Publication

Title: Small island and small destination tourism : overcoming the smallness barrier for economic growth and tourism competitiveness / Robertico Croes, PhD.
Names: Croes, Robertico R., author.
Series: Advances in hospitality and tourism book series.
Description: First edition. | Series statement: Advances in hospitality and tourism book series | Includes bibliographical references and index.
Identifiers: Canadiana (print) 20210393408 | Canadiana (ebook) 20210393416 | ISBN 9781774637234 (hardcover) | ISBN 9781774637357 (softcover) | ISBN 9781003277477 (ebook)
Subjects: LCSH: Tourism. | LCSH: Islands—Economic conditions. | LCSH: Economic development.
Classification: LCC G155.A1 C76 2022 | DDC 338.4/79109142—dc23

Library of Congress Cataloging-in-Publication Data

....

CIP data on file with US Library of Congress

....

ISBN: 978-1-77463-723-4 (hbk)
ISBN: 978-1-77463-735-7 (pbk)
ISBN: 978-1-00327-747-7 (ebk)

DEDICATION

To Suzette, Deaxo, and Kyle

ABOUT THE ADVANCES IN HOSPITALITY AND TOURISM BOOK SERIES

Editor-in-Chief:
Mahmood A. Khan, PhD
Professor, Department of Hospitality and Tourism Management,
Pamplin College of Business,
Virginia Polytechnic Institute and State University,
Falls Church, Virginia, USA
Email: mahmood@vt.edu

This series reports on research developments and advances in the rapidly growing area of hospitality and tourism. Each volume in this series presents state-of-the-art information on a specialized topic of current interest. These one-of-a-kind publications are valuable resources for academia as well as for professionals in the industrial sector.

BOOKS IN THE SERIES:
Food Safety: Researching the Hazard in Hazardous Foods
Editors: Barbara Almanza, PhD, RD, and Richard Ghiselli, PhD

Strategic Winery Tourism and Management: Building Competitive Winery Tourism and Winery Management Strategy
Editor: Kyuho Lee, PhD

Sustainability, Social Responsibility and Innovations in the Hospitality Industry
Editor: H. G. Parsa, PhD
Consulting Editor: Vivaja "Vi" Narapareddy, PhD
Associate Editors: SooCheong (Shawn) Jang, PhD,
Marival Segarra-Oña, PhD, and Rachel J. C. Chen, PhD, CHE

Managing Sustainability in the Hospitality and Tourism Industry: Paradigms and Directions for the Future
Editor: Vinnie Jauhari, PhD

Management Science in Hospitality and Tourism: Theory, Practice, and Applications
Editors: Muzaffer Uysal, PhD, Zvi Schwartz, PhD, and
Ercan Sirakaya-Turk, PhD

Tourism in Central Asia: Issues and Challenges
Editors: Kemal Kantarci, PhD, Muzaffer Uysal, PhD, and Vincent Magnini, PhD

Poverty Alleviation through Tourism Development: A Comprehensive and Integrated Approach
Robertico Croes, PhD, and Manuel Rivera, PhD

Chinese Outbound Tourism 2.0
Editor: Xiang (Robert) Li, PhD

Hospitality Marketing and Consumer Behavior: Creating Memorable Experiences
Editor: Vinnie Jauhari, PhD

Women and Travel: Historical and Contemporary Perspectives
Editors: Catheryn Khoo-Lattimore, PhD, and Erica Wilson, PhD

Wilderness of Wildlife Tourism
Editor: Johra Kayeser Fatima, PhD

Medical Tourism and Wellness: Hospitality Bridging Healthcare (H₂H)©
Editor: Frederick J. DeMicco, PhD, RD

Sustainable Viticulture: The Vines and Wines of Burgundy
Claude Chapuis

The Indian Hospitality Industry: Dynamics and Future Trends
Editors: Sandeep Munjal and Sudhanshu Bhushan

Tourism Development and Destination Branding through Content Marketing Strategies and Social Media
Editor: Anukrati Sharma, PhD

Evolving Paradigms in Tourism and Hospitality in Developing Countries: A Case Study of India
Editors: Bindi Varghese, PhD

The Hospitality and Tourism Industry in China: New Growth, Trends, and Developments
Editors: Jinlin Zhao, PhD

Labor in Tourism and Hospitality Industry: Skills, Ethics, Issues, and Rights
Abdallah M. Elshaer, PhD, and Asmaa M. Marzouk, PhD

Sustainable Tourism Development: Futuristic Approaches
Editor: Anukrati Sharma, PhD

Tourism in Turkey: A Comprehensive Overview and Analysis for Sustainable Alternative Tourism
Editor: Ahmet Salih İKİZ

Capacity Building Through Heritage Tourism: An International Perspective
Editor: Surabhi Srivastava, PhD

Medical Travel Brand Management: Success Strategies for Hospitality Bridging Healthcare (H2H)
Editors: Frederick J. DeMicco, PhD, RD, and Ali A. Poorani, PhD

Strategic Management for the Hospitality and Tourism Industry: Developing a Competitive Advantage
Editor: Vincent Sabourin, PhD

Tourist Behavior: Past, Present, and Future
Editors: Narendra Kumar, PhD, Bruno Barbosa Sousa, PhD, and Swati Sharma, PhD

The Hospitality and Tourism Industry in ASEAN and East Asian Destinations: New Growth, Trends, and Developments
Editors: Jinlin Zhao, PhD, Lianping Ren, D.HTM, and Xiangping Li, PhD

Event Tourism in Asian Countries: Challenges and Prospects
Editor: Shruti Arora, PhD

ABOUT THE SERIES EDITOR

Mahmood A. Khan, PhD, is a Professor in the Department of Hospitality and Tourism Management, Pamplin College of Business at Virginia Tech's National Capital Region campus. He has served in teaching, research, and administrative positions for the past 35 years, working at major U.S. universities. Dr. Khan is the author of several books and has traveled extensively for teaching and consulting on management issues and franchising. He has been invited by national and international corporations to serve as a speaker, keynote speaker, and seminar presenter on different topics related to franchising and services management. Dr. Khan has received the Steven Fletcher Award for his outstanding contribution to hospitality education and research. He is also a recipient of the John Wiley & Sons Award for lifetime contribution to outstanding research and scholarship; the Donald K. Tressler Award for scholarship; and the Cesar Ritz Award for scholarly contribution. He also received the Outstanding Doctoral Faculty Award from

Pamplin College of Business. He has served on the Board of Governors of the Educational Foundation of the International Franchise Association, on the Board of Directors of the Virginia Hospitality and Tourism Association, as a Trustee of the International College of Hospitality Management, and as a Trustee on the Foundation of the Hospitality Sales and Marketing Association's International Association. He is also a member of the several professional associations.

ABOUT THE AUTHOR

Robertico Croes, PhD
Professor of Tourism Economics and Management,
Rosen College of Hospitality Management,
University of Central Florida, Orlando, Florida

Robertico Croes, PhD, is a Professor of Tourism Economics and Management at the Rosen College of Hospitality Management. He currently serves as Editor of the Rosen Research Review at the University in Central Florida. According to the recent prestigious Academic Ranking of World Universities (2020 Shanghai Ranking), http://archive.shanghairanking.com/Shanghairanking-Subject-Rankings/hospitality-tourism-management.html, the Rosen College is the number one USA program.

Dr. Croes is the author of five books. His other four books are on tourism and poverty reduction (titled *Poverty Alleviation Through Tourism Development*), on the challenges facing small island destinations (titled *The Small Island Paradox, and Tourism Management in Warm-water Island Destinations*), and on the application of demand models to a small economy and is titled *Anatomy of Demand in International Tourism*. Additionally, he is a contributor to more than a dozen books. He issued more than 100 published works and emitted more than 30 industry reports.

His research has also been presented in the predominant industry and research conferences around the world, including South Africa, Hong Kong, Ireland, Turkey, Taiwan, Trinidad, and Tobago. His fields of interest include econometrics applications in hospitality, tourism demand analysis/forecasting, tourism economic impact, competitive, and sustainable tourism in tourism development analysis, tourism development applied to poverty alleviation, and tourism development in developing countries.

Dr. Croes received his doctorate from the University of Twente, the Netherlands. His dissertation focused on quantitative modeling of tourism demand, tourism development, and government intervention. He has published various articles in such journals as the *Annals of Tourism Research, Journal of Travel Research, Tourism Economics, International Journal of Tourism Research, Tourism Management,* and *International Journal of Hospitality Management.* He has served on the editorial board of 11 journals. Dr. Croes is the recipient of the 2018 and 2015 Thea Sinclair Award, the 2015 UCF Research Incentive Award (RIA), 2015 Best Graduate Student Research Paper, presented at the 32nd International Association of Hospitality Financial Management Education Research Symposium. New York University, N.Y., November 7, 2015, and several other awards.

CONTENTS

ABBREVIATIONS

2SLS	two-stage least squares
ACP	African Caribbean Pacific
B	bureaucracy
CBI	Caribbean Basin Initiative
CGE	computer general equilibrium
CTO	Caribbean Tourism Organization
CV	coefficient of variation
ELG	export-led-growth
EU	European Union
GDP	gross domestic product
HDI	human development index
HP	Hodrick-Prescott
IQR	interquartile range
MES	minimum efficient scale
MI	migration
R	remittance
SIDS	small island developing states
SMOP	SMallness and OPenness
SNIJ	subnational island jurisdictions
TALC	tourism area life cycle
TLG	tourism-led growth
TLGH	tourism-led growth hypothesis
ToT	terms of trade
TS	tourism specialization
UNCTAD	United Nations Conference on Trade and Development
UNDP	United Nations Development Program
WEF	World Economic Forum
WTO	World Tourism Organization

FOREWORD

There are few better qualified to pen such a text as this on small islands and other similar small destinations than Robertico Croes. "Tico," as he is known among his colleagues at Rosen College, University of Central Florida, has a unique history of combining high political office with a distinguished academic career while still sustaining through research and consultancy his contacts with the Caribbean and Latin America. From 1980 to 1990, he first followed a career in Foreign Affairs before eventually becoming a Minister of initially Economic Affairs and Tourism to then becoming the Finance Minister of Aruba in the Dutch Antilles from 1993 to 2001. For almost four years, from 1995 to 1998, he was the Vice-President of the Caribbean Tourism Organization (CTO), and those experiences informed much of his subsequent research.

It was probably soon after a move that, in my capacity as editor of the academic journal, *Tourism Management,* I began to receive manuscripts from Tico as a new voice representing the development of tourism in the Caribbean, writing first of his beloved Aruba. In return, I began making requests of him as a referee, and it became quickly evident that he possessed both practical experience and the technical skills to make informed opinions about the economics, finance, and social implications of tourism.

Those skills are to the fore in this book, and as readers, we are fortunate in having such an expert to guide us through the paradoxical and complex relationships as to how small destinations, often perceived by tourists as an escape from a globalized world into a retreat of peaceful land- and seascapes, can also be full members of the developed world, making as much social and economic progress as their larger neighbors. Tico begins his analysis with another duality, that of openness as against being closed to outside influences, and early in the book establishes that being small, but open, does not inhibit economic growth. Indeed, it encourages economic development.

Growth, however, is but an interim objective and not a goal in itself. The purpose of both tourism and economic growth is to create benefits for human society, and in Chapters 3 and 4, Tico addresses the issues of leakages and angst that arise from tourism in smaller destinations as they seek to create net gains, sometimes from positions of relative economic weakness. However, in the modernity of the developed world, small destinations are, it is argued, a luxury good for which there is a high demand, and given this, Tico considers in some detail the tourism-led growth hypothesis.

Given his political background in Aruba, those interested in the power play of politics may find Chapter 6 of interest, especially in his comments about the success of Aruba in its resistance to the demands of The Hague. Equally telling is his analysis of how small island states can mitigate exogenous shocks and not experience the volatility of larger economies. The final two chapters assess how small destinations can continue to create successful economies. Tico is unhesitating in not avoiding controversial statements, and Chapter 7 subjects the tourism-led growth hypothesis to several analyzes—at one stage, suggesting the notion is "foggy" (p. 178); but such statements prompt thought and are subjected to a thorough analysis.

At the same time, while engaged in technical issues, the book is written in an easily accessible style, and the non-economists will find it easy enough to negotiate their way through the various arguments. Hence it is not surprising for one engaged in empirical analysis that Tico, in the last chapter, emphasizes the importance of data and its collection, and indeed the appropriateness of data analytical techniques when summarizing and concluding the lessons learned from his research and practical research. It is a pleasure to recommend this book to its readers.

—**Chris Ryan**
University of Waikato, Hamilton, New Zealand

ACKNOWLEDGMENTS

I started writing this book about two years ago. In the beginning, the idea was to collage my research of 15 years about small island destinations and focus in particular on my fascination with the development accomplishment of my native island. During the journey, I tweaked my script to involve my thinking about the power of TS and the opportunities tourism provides to small islands to face head-on the scale issues that confound these destinations. The TS hypothesis results from changing my theoretical compass and pivoting from David Ricardo to Adam Smith to discover the journey to scaling up.

Throughout this journey, I have gotten the help of so many people with their comments, intellectual insights, and support. That said, a special thanks to my PhD students, especially to Wen Zhang, and my co-authors in multiple articles about small islands, Manuel Vanegas, Jorge Ridderstaat, Manuel Rivera, Kelly Semrad, Yang-Yang, Noah Mueller, Mathilda van Niekerk, Peter Nijkamp, Eric Olson, and Seung Hyun Lee. The articles we wrote together formed a large part of this book. I am particularly indebted to Jorge Ridderstaat, who sparred with me on many occasions and provided thoughtful comments and ideas to strengthen several chapters. A special thank you to Stephen Pratt, Eduardo Parra Lopez, Andrew Spencer, and Alan Fyall, who delivered meaningful words to the book's first two chapters and a blurb for the book's cover. Many thanks to Dapeng Zhang for his insightful comments. I am very much indebted to Luiza Semrad, who spent many hours perusing and editing each of the chapters. I enjoyed and benefited enormously from her patience, creativity, and editing insights in the writing of the book.

The pandemic hit the world almost at the end of this book journey, forcing me to revisit small island destinations' resilience. I participated in two webinars that helped me deepen my understanding of the islands' resilience, particularly in the Caribbean region. I am indebted to the University of the West Indies, particularly Michelle McLeod, who hooked me up with some brilliant minds and practitioners in the Caribbean. My special thanks to Vincent Vander Pool Wallace, the former Minister of Tourism in The Bahamas. Jafar Jafari and Reza Soltani organized the other webinar.

They were gracious enough to involve me in a lively conversation with members of the community government initiatives in Jamaica and the Caribbean Diaspora, discussing how small islands could pivot to a new state of conducting business with novel strategies and business concepts, partners, markets, and products.

I am also indebted to my eldest son, Deaxo, who was gracious enough to help assess the response of small Aruban firms to the pandemic and for scheduling meetings with some local bankers. I was impressed by what I learned from small businesses bouncing back and bouncing forward opportunities within small islands.

Finally, a big thank you to my amazing wife, who always believed and encouraged me to complete this book.

ENDORSEMENTS

"Although often small in size, islands make a disproportionately large contribution to international tourism, where policies of economic specialization have endured for decades. Authored by an "islander" with both political and academic acumen, this timely contribution offers a thoughtful, engaging, and provocative insight into the future of islands and the means by which the tourism specialization hypothesis will continue to serve as the platform for their future competitiveness."

—Dr. Alan Fyall, Visit Orlando Endowed Chair of Tourism Marketing, Rosen College of Hospitality Management, University of Central Florida

"Dr. Robertico Croes' new book on tourism specialization seeks to explore the relationship between smallness, tourism specialization, economic growth, and tourism competitiveness among small island destinations. The topic of small islands has not received a great deal of attention in academic literature. The academic literature that does exist tends to be overly simplistic and reductionist (there are inadequate resources and limited opportunities for inhabitants that may be inadequate, inept, and too ineffectual to control their own destiny). Dr. Croes takes a systematic and rigorous approach to ask, is 'smallness' in and of itself necessarily a problem? Croes explores the definition, measurement, and concepts of smallness to question whether a country's small size impedes economic growth. To answer this question, Dr. Croes considers existing theories of Ricardo's competitive advantage and Adam Smith's dynamic competitive advantage. After defining specialization, Croes puts forward his tourism specialization hypothesis, noting it is a dynamic process, bringing together supply and demand, societal learning, and openness to overcome any small barriers in achieving economic growth and prosperity. This volume by Dr. Croes is one of the first attempts to move beyond the descriptive analysis of tourism in small islands by developing a tourism specialization hypothesis based on a solid, rigorous review and extension of existing applicable economic theory to better understand the relationship between tourism in small island destinations and the wider

socio-economic environment. This book will challenge your thinking and challenge some of your existing thinking in this area."

—Dr. Stephen Pratt, Professor and Head of School, School of Tourism and Hospitality Management, The University of the South Pacific, Headquarters and Laucala Campus

"Dr. Robertico Croes is a professor at Rosen College (USA), an expert in tourism economics, human development, poverty, and destination management with a special interest in islands and developing economies. His experience in recent years has allowed him to have a critical vision, but at the same time reflective and deep strategic vision, of how the islands should approach their developments and competitiveness. The book proposes an acute and intelligent approach on the tourism specialization (so necessary and scarce in the literature) in the islands, supported by the base of eight chapters, which analyze the reality of these territories and their needs, offering own vision on the subject, although always in a reasoned and consistent way. It starts with a conceptualization and extensive revision of the construct "smallness," advancing on the hypotheses of his impulse and how to achieve it in a context of global changes and uncertainties (Chapters 1 and 2). In a collation of these first chapters, the author compiles a variety of aspects that serve to build the discourse of specialization and its scope and how all this can lead us towards a strategic optimization of benefits. Chapters 4–6 give a sustained analysis of the reality of tourism specialization and economic growth, based on size, market opening, and a comparative evaluation of destinations in the Caribbean. To close with an analytical approach of the heterogeneous levels of work and policies on which the various approaches to specialization must be addressed. The reader of this magnificent book, has an agile and entertaining critique, with the possibility of educating in the analysis of island spaces and their developments."

—Dr. Eduardo Parra López, Professor of Digital Economy, Organization of Companies and Tourism, Professor of the Chair of Tourism of the University of La Laguna, and President of the Spanish Association of Scientific Experts in Tourism (AECIT)

"I am excited about this text by Professor Robertico Croes: "Small Island and Small Destination Tourism." This text excavates and ignites the latency in knowledge creation on such a critical subject. The book appeals to me through three lenses: (1) As a tourism academic; (2) As the former CEO of Jamaica's Tourism Development Agency; and (3) As a national living in a small island developing state (SIDS). From these perspectives, Croes provokes deep thought about a myriad of issues and provides a call to action in the midst of complexity. He does this in a way that demystifies nuances related to SIDS as he skillfully weaves together theory and praxis that will cause SIDS to recognize that no longer can their physical size be allowed to dwarf the enormity of their thinking. It is a line by line, page by page riveting read!"

—Andrew Spencer, PhD, Professor of Tourism Development,
University of the West Indies, Mona, Jamaica

PREFACE

This book is a product of my longstanding interest in understanding the nature of small island destinations' economic development and specialization. For a very long time, our thinking pertaining to development anchored itself around three main ideas: vulnerability, dependence, and lack of growth. History associated the latter with an existential threat to these small islands. In this book, I principally strive to illuminate the potential power signified in the small island destination as a stable and enduring tourism entity. But, in its sentences and between its lines, I also strive to leave the reader with some understanding of the power embodied in the people who call the small island home-for without their manifest spirit, their coherent and constant struggle to be, to reach for a means to gain prosperous well-being, the small island would certainly be burdened beyond measure. The character of a people can go a long way toward realizing, generating, and securing opportunities that could lead to meaningful life achievements born of the economic growth their steadfast character grants. Arthur Lewis, in his seminal work *The Theory of Economic Growth*, aptly describes the influence of human behavior on economic growth.

> *Given [a] country's resources, its rate of growth is determined by human behavior and human institutions: by such things as the energy of mind, the attitude toward material things, willingness to save and invest productively, or the freedom and flexibility of institutions. Natural resources determine the course of development, the pattern, and the shape of the development process, but it is behavior and institutions that determine growth (p. 53).*

At its core, this is a book about smallness, tourism specialization (TS), economic growth and tourism competitiveness: in particular, small islands practicing warm water island tourism. Small islands were not always popular subjects in social sciences. Often neglected, omitted, or considered obscure, inquiry regarding small islands was scarce. Baldacchino claims that islands are often considered insignificant inquiry matter. Also, social science literature has mainly considered small islands as unimportant, its people and governing structures as takers from or beggars of the outside world, with little agency to determine and shape their own future. The populations of these islands have been looked upon as passive, with scarce creativity to discern developmental pathways that could realize prosperity and achieve a decent and consistent standard of living over time. However, nothing could be less truthful. To characterize small island people as takers or beggars consigns them to a useless status-inadequate, inept, and too ineffectual to control their own destiny. Instead, small island people often boast the need and desire to be self-sustaining, are capable of fostering the motivation to succeed, and possess the creative spirit to cultivate an enduring way of life. This is a characterization closer to the truth. This insight augmented by the natural magnetism of the islands' geographical appeal affords a more affirmative footing upon which to build an industry in service to the islands and their populations.

Such islands seem to have a special allure to peoples' minds and fantasies. Butler refers to this allure as the "Robinson Crusoe factor." Others, such as Scheyvens and Momson, posit that the very smallness and isolation of islands reveal and entice much of what the paradise concept epitomizes. Islands entertain something mysterious and magical, feared but wanted, and, far more often than not, appealing to mankind as a place to live. Powerful metaphors have referenced the idea of an island, depicting the very essence of Western culture. Homer, St. Augustine, Shakespeare, More, Daniel Defoe, Mark Twain, James Joyce, Camus, George Orwell, and Eco, among others, enthralled the mind with the magical ambience of insular mythical realities to escape religion, ideologies, de-stress daily doldrums and routines, search for renewal or self-discovery, or simply to tickle their fantasy about paradise as an alternative society.

The overall concept of travel is given to both unique and general characterizations. Some travel for escape, some for experience, some for education, some for searching, some just because travel is something to do that almost everyone else does, and on and on. Perhaps more than any other tourist avenue, small islands boast for the traveler a place to go for peace and relaxation (including landscape, beaches, and people)-a place where stress and the confines of the work-a-day world in which a tourist must live do not exist. Here, the traveler is not "traveling" in the sense of the word. Instead, the traveler is going on a journey that involves his own existence in commune with only and all of that which the traveler imagines contributes to a simple, uncomplicated being-a time just to be, a time to be a part of the faraway mystique that by its nature mentor's serenity, reflections of the mind, body, and soul on that which is worthwhile in the world. If such a place offers this opportunity and more, it must be marketable.

In spite of this, questions still loom. Why is there often the expectation that visiting small islands means walking into a time of little to no progress in its society, a backwards place where dirt roads, hovels, unforgiving heat, and a people with little to no care about their subsistence live day by day; where quality of life is just a phrase-an unattainable circumstance. Perhaps, in the traveler's interpretation of peace, she requires a place untouched by any part of the progressive society she has left. If the traveler cannot experience his interpretations and characterizations of peace, he may as well have stayed home. But, why is there the confidence and conviction that small islands cannot progress? Literature has perhaps been lax in its research regarding perspectives on the prosperity and ongoing development that small islands could achieve. It has perhaps too often relegated the small island destination as a place without the capacity to grow in a manner that preserves its tranquility while forwarding its development with the means to sustain that development and its positive economic result.

By the end of the 20th century, I noticed a brewing of interest and proliferation of articles regarding small islands. Just in the first decade of the 21st century, several journals dedicated a complete issue to the study of small islands including Tijdschrift voor Economishe en Sociale Geografie, Geographical Review and Journal of Development Studies. Also, the Island Studies Journal was launched in 2006. The interest of the social sciences to investigate small islands seems due to globalization of the world economy, personal mobility, and affordable technology. International organizations also dedicated attention to the small island problematic. Issues and challenges related to the small island development quandary were etched in the Barbados Program of Action in 1994 and reiterated and reaffirmed in the Mauritius Declaration of 2005. The European Union (EU) also revealed heightened attention to the small islands question. These forces conjured making small islands accessible to researching the potentiality for tourism as a prosperous entity.

References as to two main limitations confronted by small island tourism development have been made. One is the small islands' susceptibility to their potential for catastrophic weather disasters, as well as those disasters perpetrated by man (terrorist activity, poor governance, etc.). Another is the limited opportunity to perpetuate economies of scale, as well as specialization: this being the consequence of the availability of adequate or sufficient resources with which to build and maintain a lucrative national market of goods and services. Moreover, research has indicated small islands to be strongly suspect in their ability to overcome size, resource availability, and disaster constraints in order to prosper and continue to do so in a sustainable manner. However, the effects of these difficulties on growth seem inconclusive. For example, in their extensive empirical review, Easterly and Kraay demonstrate the economic results of small countries were as good as their larger neighbors. Other studies, such as Brau, revealed that small countries experienced high growth rates despite the difficulties they faced. The implication is that size seems not to matter.

These concerns as revealed in the mainstream literature prompted my interest in examining the deterministic and fatalistic view regarding small islands' viability, and to discern whether small islands had options to grow and achieve prosperity. By prosperity, I mean, achieving life conditions that enable a flourishing existence. This characterization of prosperity follows Sen's assertion that prosperity relates to capabilities to flourish and to lead the life that one values. Of course, this view encompasses material conditions (such as jobs and income) but also other aspects (e.g., inclusion, health, our family happiness, the strength of our relationships and our trust in the community, and satisfaction at work and our sense of shared meaning and purpose) of quality of life.

There are many questions revolving around small island prosperity and the means by which a success above risk status could be sustained. What we do see is that islands have not been successful in diversifying their economies away from tourism to mitigate their risks. They have not been able to experience the evolutionary processes and dynamics that occur on mainlands, and seem to be "stuck" to tourism development. My question is: Is tourism development the low-hanging fruit that impedes islands from reaching for higher-hanging fruits of development that other countries are experiencing? I was further consumed by several questions, such as; can small islands spur economic growth on a sustainable level, can small islands prosper and propel a decent quality of life for their residents, can tourism be a reliable and resilient development strategy for small islands, is the Sun, Sand, and Sea (SSS) as a tourism development model competitive, and will the model persist? Especially, does size automatically relegate a small island region to negligible development prospects-the preponderance of which could contribute to fatalistic views; or, is there a path to prosperity to challenge such perceptions? These questions directed my research focus in a systematic way and characterized my early scholarly research during the first decade of this century.

The chapters discussed in this book delve into the aforementioned questions. In exploring these questions, I applied the duality technique, so my work focuses on particular contexts while simultaneously considering generality. To achieve duality, there is need for data collection while addressing the imperative to show the ability to transcend these data to develop a theoretical understanding. Without a system of interpretation, which is what theory is, data is no more than interesting observations that imply no conclusions for tourism development theory.

My early scholarly work embarked on understanding the processes involved in a newly developing tourism industry and the role of tourism in triggering these growth patterns. Essentially, these aspects refer to a very important component that small island destinations seem to have overlooked, i.e., the need to undergird decision-making via smart framing and rigorous research. This, and my questions, led me to primarily support my early research on four concepts: smallness, tourism specialization, economic growth, and tourism competitiveness. The tourism specialization construct is related to tourism supply and demand, while the tourism competitiveness construct relates to tourism demand. The concept of smallness relates deeply to the ability of small islands to strategize a prosperous road to sustainable economic growth through tourism.

My inquiry through my writings adheres to the method suggested by Karl Popper. He asserts that (imperfect) observations of the world are the basis of all sciences, and that the scientific method rests on inferring general laws from these observations, supported by further (critical) testing (falsifying hypothesis). Popperian logic in the tourism context means that a deductive process between theory and evidence is never aimed to generate certainty, but to gradually increase our understanding of the tourist phenomena under study.

In searching for meaningful insights, I continued to fundamentally mold my early research on smallness, tourism specialization, and tourism competitiveness. When and wherever, there was need, I expanded the original writings, albeit without removing the essence of my early conceptualization and understanding of tourism development theory. This book is testament to small islands' ability to overcome the constraints that literature has so often reported as too daunting a challenge to overcome. It strives to reveal tourism as a potential and powerful pathway, mediated by contingencies, to prosperity. The findings give a sense of empowerment that can help small islands endeavor to improve the lives of their residents. Hope and an optimistic look at opportunities for small islands can lead to remarkable life improvements and may serve as a powerful incentive to undertake efforts towards prosperity.

This book reveals my views, professional assessments and interpretations, and first-hand experiences with development of small island destination development and management. Moreover, in my professional career I have endeavored to extensively research theories and hypotheses with regard to the tourism industry as a means by which small island destinations could develop and sustain economic prosperity. Thus, while this book is written in a manner that reveals the methods and procedures required of professional research, its professional foundation is shared with my ardent understanding that, while the destination may be small, it is a mistake to presume that the creative force of its people bears an equivalent smallness with which to contour their own life course and well-being.

So, the nature of this book is to provide witness to the promise that small islands possess toward a competitive legitimacy in the global tourism market. However, I hold an underlying passion regarding the value of a peoples' creative agency to forge a quality of life and well-being beyond the impressions of outside expectations. I have seen and experienced the significance of constructive thought and action harnessed by the creative mind, of adaptation to traumatic forces without which destructive consequences could emerge, of the pact that people share between their station and their responsibility to overcome their problems, and of their dexterous resourcefulness and resilience in their journey to adapt, change, and succeed.

As a last note, while the need to corral creative thought is burdensome to all, it cannot be assumed to be absent. In short, we cannot equate the size and value of a peoples' creative vitality as they pursue their road to prosperity with the geographical size of their island. Indeed, it is often more likely that the immeasurable constant of creative vitality often transcends the measurable topography.

By the time I finished writing the book, global tourism had an unexpected guest that marred mobility and social interaction, the main twin arteries of fueling tourism. COVID-19, stoking social distancing and forcing people to stay at home, is unable to coexist with tourism. In the last chapter of the book, I will assess the impact of COVID-19 on the tourism specialization hypothesis and its ramifications for small island destinations. Whether small island populations can rise from beneath the COVID-19 burden that threatens their economic future is yet to be seen. However, given that such populations have risen from desperate circumstances in desperate times, it is and must be entirely possible.

"Every small nation has some advantage in natural resources—whether it be location, coastline, minerals, forests, etc. But some show capacity to build on it, if only as a starting-point, toward a process of sustained economic growth and others do not. The crucial variables are elsewhere, and they must be sought in the nation's social and economic institutions."

—Kuznets (1960)

"The insights of endogenous growth theory suggest that small states are well-placed to enjoy relatively high rates of growth, in spite of their economic sub-optimality, because of their high degree of openness to trade and propensity for capital formation."

—Read (2002)

"I am my choices. I cannot not choose. If I do not choose, that is still a choice. If faced with inevitable circumstances, we still choose how we are in those circumstances."

—Jean-Paul Sartre (1946)

CHAPTER 1

RETHINKING SMALLNESS AND OPENNESS

The interest in defining an optimal country size to overcome "the penalties of smallness" intermittently piqued the attention of philosophers and scientists alike even as far back as Ancient Greece. For example, during the American Revolution, the founding fathers struggled to justify the large size of the country as a benefit to economic prosperity. Thus, the years of perplexity and pondering have made clear that the geographical size of a country and its relationship to prosperity bears an importance. Yet, the literature on the economic and social implications of the size of countries is scarce. The editorial of *The Round Table* (*The Commonwealth Journal of International Affairs*) succinctly reveals the development literature's paucity regarding the effects of country size, stating "Academia has paid little attention to small states" (2012, p. 202). This lack of attention continues despite the proliferation of small states and countries into the global environment.

The proliferation of small countries in the global arena is due to the significant decrease in the economic costs of independence occurring since World War II. The ever hospitable open global, trading order enabled small countries to prosper within a more secure global system. These two trends, i.e., the paucity in examining the implications of small size and their global proliferation, have pitted centrifugal against centripetal economic and social forces about their viability and role in the new global system. It is likely critical, then, to examine the impact of smallness (size) on the economic and social potential of countries-particularly since their welfare may well depend on the small-scale problems that could impede their sustained progress toward that potential.

1.1   THE CHANGING MEANING OF SMALLNESS

The literature examining small countries describes them as tethered to catastrophe. Typically, smallness references material resources, constraints, and distinct characteristics measured in a geographic space. The geographic space involves states in the sense of international politics and relations, territories, or the subnational island jurisdictions (SNIJ), including Aruba, Cayman Islands, and Bermuda. The latter concept is borrowed from Godfrey Baldacchino.[1] The literature assessing these small countries identified them as non-viable, vulnerable, weak, and insignificant entities in the global environment. This non-flattering, albeit defeatist, characterization seems to emanate from the conceptualization of smallness through a resource-based lens. Moreover, possessing resources determines the opportunities or limitations available for a country to prosper and to realize its inclinations in the global setting. That is, the presence or absence of water, beaches, deserts, mountains, oil, and gas can have an influence on how people behave and interact with each other, and ultimately how they organize and coordinate on an intra and inter-organizational basis to achieve collective goals. That available resources have shaped and molded the evolution of societies and civilization is evident.

One can then understand why debates about the meaning and relevance of smallness goes back more than two millennia ago. From early civilizations until the French Revolution, commentators saw smallness as a beneficial attribute for survival and prosperity for a collectivity. For example, Plato and Aristotle touted the societal benefits that could be derived. Plato estimated that the optimal population size of the polity to thrive should be a maximum of 5,040 heads of families. Aristotle considered the optimal size to be when everybody knows everybody. The desirability of a small polity to prompt prosperity and inclusiveness also resonated with Montesquieu and Rousseau. They embraced an optimal population as a small Greek polity. For example, Montesquieu asserts,

> *It is in the nature of a republic that it should have a small territory; without that, it could scarcely exist. In a large republic, there are large fortunes, and consequently little moderation of spirit... In a large republic, the common good is sacrificed to 1,000 considerations; it is subordinated*

---

[1] Baldacchino addressed the smallness problematique in multiple writings. See some of his writings in the reference list.

to various exceptions; it depends on accidents. In a small republic, the public good is more strongly felt, better known, and closer to each citizen... (From The Spirit of Laws, C.L. Montesquieu, 1750, Book VIII).

The enlightenment thinkers view smallness as the anchor for freedom, attachment to the public interest, and the foundation for a sustainable republic. However, the advocacy for smallness as a virtue changed fortune with the American Revolution. The founding fathers strenuously debated whether smallness was tenable for the American Republic. They came to the opposite conclusion that they should advocate for largeness. For example, Hamilton questioned Montesquieu's view about the relevance of smallness in defining the republic.

If we, therefore, take his [Montesquieu] ideas on this point as the criterion of truth, we shall be driven to the alternative either of taking refuge at once in the arms of monarchy, or of splitting ourselves into an infinity of little, jealous, clashing, tumultuous commonwealths, the wretched nurseries of unceasing discord and the miserable objects of universal pity and contempt (Hamilton, The Federalist No. 9).

This view reveals the fear that smallness would create instability and weakness. Mills, who also viewed smallness as a temptation to war and foreign aggressiveness, later adopted this view. Not only was small-ness considered as a perversion of the democratic and peaceful ideal, but smallness also engendered a sense of limitations, constraints, and vulner-ability. This negative connotation persisted during the establishment of the League of Nations when smallness impeded Liechtenstein from attaining membership in that organization.

The notion of equating smallness with vulnerability emerged promi-nently after the Second World War. Already, in September 1957, the Inter-national Economic Association organized a conference in The Hague to examine the economic implications of the small size of nations in the global setting. The conference *The Economic Consequences of the Size of Nations*, in particular, assessed whether small size can inhibit the viability of these nations in the global setting. The conference published its proceed-ings in 1960. Robinson, in the introduction of that publication, stated that the study of the economic implications of small nations is a "subject that well deserves more attention."[2] This quote suggests that the attention to

---

[2] See Robinson (1960).

this subject lacked currency at the end of the 1950s. Indeed, the conference's attentions centered on the advantages of scale embedded in large countries and whether small nation's international trade could overcome "the penalties of smallness."

However, this important issue did not enjoy the hoped-for attention of the 1957 conference nor of the years between. The topic was addressed only sporadically and intermittently with its concomitant vulnerability characterization. For example, William Demas expressed in connection with the Caribbean islands: "Small may be beautiful, but it may also be fragile, vulnerable, and extremely externally dependent."[3] The focus on smallness and its concomitant economic consequences reemerged during the 1980s for several reasons, including the heightened attention paid by the World Bank and other international organizations, the creation of the Commonwealth Vulnerability Index, the persistent calls of special status and deferential treatment of small countries within the World Trade Organization, and the formation of a Consultative Group of Small Economies within the negotiations' framework of the Free Trade Area of the Americas.

Some sporadic attempts appeared among the United Nations units, such as the initiative in 1972 of the United Nations Conference on Trade and Development (UNCTAD). UNCTAD prompted a program, the so-called small island developing states (SIDS). This program distinguishes the relevance of smallness as an essential criterion to define countries. Vulnerability became the key criterion to define smallness. The work of the Commonwealth Secretariat in the 1980s followed this line of thinking again, expressing vulnerability in defining smallness. Vulnerability is the lack of resources of a country to provide for itself-depicting helplessness, particularly in economic terms. The premise of these initiatives or actions is that small economies are a distinct group of countries, that these countries are vulnerable and, therefore, require special attention and actions.[4]

Two salient aspects remain constant about the characterization of smallness during the postwar period, i.e., paucity of a consistent conceptualization or measurement of smallness, and smallness as synonymous with vulnerability. The vulnerability argument mainly considers the

---

[3] See William Demas (1992).
[4] See, for example, https://www.unwto.org/sustainable-development/small-islands-developing-states; https://www.unccd.int/publications/land-degradation-neutrality-small-island-developing-statestechnical-report; and https://www.unenvironment.org/resources/report/emerging-issues-small-island-developing-states.

disadvantageous bearing imposed by small size on these countries due to external shocks. These external shocks stem from their economic openness and their susceptibility to natural disasters. Smallness prompts the need to import many goods because the demand for consumer goods is much higher than what can be produced at home. Opportunities to invest in these countries are too risky because they have limited factor endowment, and they are too open to the outside world, exposing themselves to high and substantial risk from external shocks. The result is that export receipts must pay for the high import costs. Their vulnerability ensues from the concentration on a few foreign markets, which may make small countries prone to imported business cycles[5]. Thus, these countries suffer from higher-income volatility due to their openness. Natural disasters happen with high frequency and intensity in small countries. For example, for the past 40 years, the Caribbean experienced over 250 natural disasters, inflicting substantial human and economic costs.[6]

Because of all these limitations and constraints, commentators often described small islands in bleak terms such as vulnerability, problems, and dangers. This vulnerability is at the heart of doubts expressed in the viability of smallness as a believable and dependable depiction of market capacity that large countries enjoy. Doubts about the economic viability of tiny countries include the overwhelming significance of economies of large scale in terms of administration, national markets, and the scope for the application of efficient capital-intensive technologies.[7] Small countries also lack the full range of both human and natural resources required for an indigenously motivated pattern of economic development.

The previous discussion reveals the ontological evolution of smallness. Initially, the focus was on supporting the ideal of freedom and democracy, but this eventually evolved into a disadvantageous concept equated with vulnerability. This evolution has clear historical markers, the concept being positive from the Ancient Greeks to the Enlightenment and eventually plummeting into a negative concept from the American Revolution to the recent past. However, the attention to smallness as topicality has not been consistent throughout time. This attention has been sporadic, intermittent, fleeting, and irregular.

---

[5] See, for example, Kuznets (1960).
[6] Acevedo (2014) estimated that the Caribbean, on average, lost one percent of GDP on an annual basis, suffered over 12,000 deaths, and lost nearly US$ 20 billion in material damages.
[7] See, for example, Briguglio (1995); and Streeten (1993).

Regardless of the discussion about the characterization of smallness, Baldacchino concludes that the scholarship addressing this topicality is marginal. A quick view through Google Ngrams confirms his assessment. The search on Google Ngrams includes an interesting attention timeline for the smallness topicality. The interest in the subject matter emerged around 1942, reached a peak in 1990, lost attention till 1993, and reached its highest mark in 2005, losing its luster afterward. See Figure 1.1. A similar search was also conducted using the key words *small islands*. The result in Google Ngrams suggests that the subject matter of small islands also lost scholarly attention after 2005. The decolonization process since the 1960s prompted heightened attention to the topicality of smallness. To illustrate this point, 11 states with a population of less than 1 million became United Nations members in the 1960s, 17 in the 1970s, and eight in the 1980s.

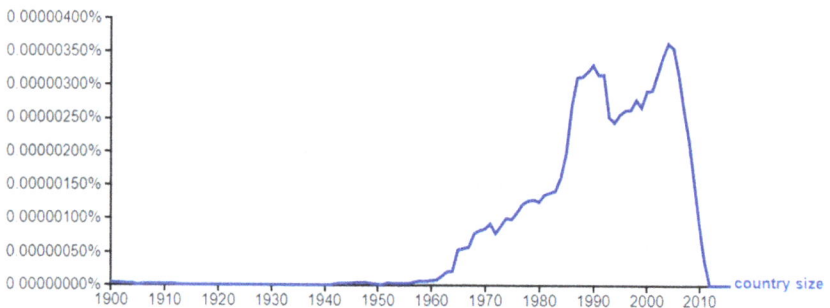

FIGURE 1.1   The evolution of country size as a subject matter.

1.2   FOUR FRAMEWORKS ON KEY DETERMINANTS ON SMALLNESS

While there is a sensible concern that smallness is not salient in the scholarship annals, what constitutes the determining factors distinguishing smallness remain an open question. Different perspectives and lenses examined the distinctiveness of smallness in particular regarding the economic and social potentials of smallness. The debate distils four distinct claims regarding the threshold size that separates a large from a small country.

1.2.1 THE SOCIOLOGICAL INTIMACY FRAMEWORK

The first claim relates to the sociological perspective espoused by Benedict in his sociological article, *Sociological characteristics of smaller territories and their implications for economic development.* Benedict's claim asserts that scale defines the intensity of personal contacts and the concentration of roles in one person. This intensity and concentration of roles engender an intimacy that limits impersonal role-relationships and reveals a high degree of personalization of decision-making. According to Benedict, "In a small-scale society, where the entire social field is small, relationships tend toward the personal pole. It matters much more who a man is than what he does." (p. 27). Limited society choice spawned by a small scale may be critical in determining the quality of governance and public institutions. Baldacchino refers to this intimacy as a prominent feature of the "small scale syndrome," and argues that social capital is a crucial determinant for economic development.

The literature that examines the small country development adminis-tration reflects the assumptions of the sociological intimacy perspective. These studies consider size as the dominant factor in shaping the problems and advantages of the tasks' administration in small countries. Person-alism and particularism seem to shape the quality of the small countries' administrations. This seeming assessment of these administrations is inconsistent with the prescriptions of the "rational-impersonal" bureau-cratic view of governance. Governance and administrative practices are often anchored on the high-stress levels occasioned by decision-makers given that they may personally know people affected by the choices and decisions they make. This stream of study suggests that any population threshold should reveal the transition point at which decision-making no longer favors or affects the decision-maker's acquaintances, and where their choices and decisions are entirely impersonal.[8]

However, the notion that country smallness yields small problems does not correspond to anecdotal and evidentiary reality for several small islands. While there is a literature stream positing that small populations prompt more social cohesion and intimacy, and hence stronger democratic institutions, there exists another literature stream that presents an opposite perspective. According to this perspective, small islands seem enmeshed with autocratic tendencies, highly personalized decision-making, and an

---

[8] See, for example, Everest-Phillips (2014); and Schahczenski (1990).

extreme lean toward a partisanship that is imbued with personal animosi-
ties. These tendencies make decision-making in poorly endowed information settings prone to self-inflicted social and economic problems.
Concerns exist to express criticism openly because of the all-pervasive
presence of governments and bureaucracies in small islands.[9] For example,
Veenendaal examined four cases: San Marino, St. Kitts-Nevis, Seychelles,
and Palau. The study found a disparity between formal democratic institutions and a political anti-democratic reality embedded in a high political
contestation. According to Veenendaal this reality is inconsistent with the
idealized notion of harmony and social consensus resulting from social
intimacy and cohesiveness.[10] Arguably, the reality of partisan politics
and authoritarianist tendencies do not seem typical realities of smallness. Rather, these tendencies seem prompted by socio-economic realities
embedded in increased inequality in Western capitalism.[11]

1.2.2   THE RATIONAL TRADE-OFF FRAMEWORK

The second claim pertains to the economic perspective asserted by Alesina
and Spolaore. In Alesina and Spolaore (1997, 2003), country size is endogenously determined as a result of a trade-off where large countries have
economies of scale in public goods provisions and higher productivity due
to broader markets, but a greater degree of preference heterogeneity. These
authors define the problem in terms of a trade-off where small countries
display preference homogeneity and social cohesion supporting democratic participation. These democratic advantages face economic disadvantages as prompted by the small size. For example, smallness implies
higher per capita costs of non-rival public goods, a small internal market;
and they do not have enough defense against larger countries.

Dahl and Tufte (1973) likely pose the most comprehensive study of
the importance of country size and is one of few studies that consider the
country's geographic area as a potential determinant of economic outcomes.
The latter can inhibit the implementation of excellent and growth-enhancing
policies and provoke domestic discontent. Domestic discontent could
lead to secessions. Alternatively, a small country may have an advantage

---

[9] See, for example, Brown (2010).
[10] See Veenendaal (2013).
[11] See, for example, Stiglitz (2012); and Mazzucato (2018).

due to its preference homogeneity. This characterization increases its democratization likelihood and stability. Alesina and Spolaore's trade-off analysis may explain boundaries evolution in many countries including the break-up of the West Indies Federation, Malaysia-Singapore, Gilbert, and Ellice Islands, St. Kitts-Nevis-Anguilla, and the Netherlands Antilles.[12] However, this premise that country size is endogenously determined may not always apply in the context of all former colonies. The reason is that the colonial powers exogenously determined their borders. This observation is particularly the case in Africa, where borders have mostly been drawn arbitrarily.

1.2.3   THE EFFICIENCY SUB-OPTIMAL FRAMEWORK

The third claim regarding the role of country size in influencing performance is the economic sub-optimality framework. This framework stems from the works of Armstrong and Read.[13] The framework eschews the assumptions of the neoclassical analysis anchored in constant returns to scale, perfect competition, and zero transportation costs. Instead, the pervasive presence of market imperfections shapes the economic performance of these small countries. Scale impacts the unit costs of production, which are higher in small economies because of the insufficient domestic demand to achieve the minimum efficiency scale (MES) and higher input prices in the production process. Insofar as small destinations possess a limited domestic market, it then follows that competition is also limited under the umbrella of small island scale disadvantages.

That is, entrepreneurial endeavors that could form linkages or synchronizations with existing businesses do not adequately emerge given that existing business numbers may very well be stunted. Thus, the goods and services required to field a competitive industry lack in availability. Overall, the very smallness of the destination serves to thwart the ingress of businesses that might otherwise have the potential to aid in the growth of the destination. Further, limited markets resulting from the umbrella of scale disadvantages could give rise to the pricing controls and dangers of a business domain that borders on or becomes monopolistic or oligopolistic-potentially giving

---

[12] These islands were forced to be part of a larger entity than themselves to gain independence or to achieve greater political autonomy. All these islands shed these colonial attempts.
[13] See, for example, Armstrong and Read (1995, 1998, 2001).

birth to more problems. Under circumstances where small scale limitations impede proficient growth of small islands' economics, necessary costs of wants as well as basic needs of the citizenry and visitors can soar, making residential and tourism life more difficult. The availability of goods and services would simply be insufficient to support needed growth.

The situation whereupon a market operates from a sub-optimal view implies that a balance between production expenditures and goods' pricing would not correspond to meet the minimum unit costs. In this situation, businesses cannot manufacture their products and go to market at competitive prices. Moreover, if small islands are functioning at sub-optimal levels, a bleak outlook for favorable market performance becomes increasingly evident. The small market size has a substantial influence on the efficiency of factors allocation. The main economic handicap feature of small islands is that they incur a large size of the minimum required efficient scale of production compared to their small demand. This handicap has consequences on how firms operate, and the high average price of unit costs affects the profitability level of firms. But the level of minimum efficient scale (MES) relative to small demand also affects the efficient functioning of the market-distorting prices to the detriment of social welfare on people who depend on such efficiency.[14]

Briguglio empirically demonstrates that the small country size prevents economies of scale in manufacturing. This structural economic reality hinders the efficient output of many goods and services in a small country.

1.2.4   THE STATISTICAL FRAMEWORK

The fourth claim embraces the concept that population size is relative and applies several distribution methods to determine the population size threshold. Kuznets used a distribution examination to establish a population threshold of 10 million while arguing that the economic structure of small countries differs from large ones. Different economic structures render the comparison of economic development based on size untenable. A cut-off level of 3 million people, based on 'a natural break in the population size continuum,' has been employed by Armstrong and Read

---

[14] See, for example, Winters and Martin (2004) who suggest that firms located in small economies would earn negative returns, making most of their products uncompetitive in the global markets. For a similar view, see Gal (2003).

(2003). Based on this population threshold, they identified 72 small entities, which are islands or archipelagos. Others, like Crowards, applied a cluster analysis and then combined population, area, and economic parameters to determine small size. His study identified 79 small countries within a sample of 188 developing countries. Brito also applied a cluster analysis combining population and land area to characterize small size. He characterized 83 countries as small within a sample of 183 countries. Easterly and Kraay define small countries with a population of 1 million or less. The Commonwealth Secretariat (1997) uses a population size of 1.5 million people. Atkins et al. (2000) reformed this population threshold in their study regarding the Commonwealth Vulnerability Index.

The statistical framework lacks a theoretical underpinning to support the identified exact population size. The identified population threshold seems to respond to capricious considerations. Kuznets, for example, set the dividing line at 10 million, even though he thought that nations with a population below 5 million would face severe developmental constraints. These constraints relate to the size of the market that hinders economies of scale and high factors' costs. Consequently, he suggested some propositions regarding the economic implications of country size. He suggested two main propositions combining economic theory with the political economy. In his view, smallness is a handicap for a country's economic growth and economic development. However, small countries, because of their higher social cohesiveness, may seize opportunities derived from technology and economic growth. Around 50 years later, Laurent addressed Kuznets's propositions regarding the economic implications of small countries. He concludes that Kuznets's propositions at the end of the 1950s have passed the test of time, meaning that small and large nations are distinctively different in their constraints and developmental strategies (Table 1.1).[15]

---

[15] Kuznets (1960) says as follows regarding the first proposition: "It would seem from the discussion so far that small countries are under a greater handicap than large in the task of economic growth." (p. 27). Laurent (2014) construed this sentence stemming from the postulations of economic theory. The second proposition stems from the following sentence: "Small nations, because of their smaller populations and hence possibly greater homogeneity and closer internal ties, may find it easier to make the social adjustments needed to take advantage of the potentialities of modern technology and economic growth...." (p. 28). Laurent (2014) references this sentence as a manifestation of the postulations of the political economy perspective.

TABLE 1.1   Frameworks Assessing the Relevance of Size

Framework	Premise	Important Authors
The sociological intimacy framework	Limited impersonal relationships and a high degree of personalization in decision-making	Benedict (1966); Baldacchino (2012)
The rational trade-off framework	Trade-off between the scale in production, costs, and preference heterogeneity	Alesina and Spolaore (2003); Alesina (2003)
The efficiency sub-optimal framework	Small scale spawns insufficient domestic demand to achieve MES	Kuznets (1960); Briguglio (1998); Easterly and Kraay (2000)
The statistical framework	Distribution methods determine scale thresholds	Crowards (2002); Brito (2015)

However, as the economic costs to become politically independent have steadily withered over time, the country size has also decreased accordingly. That is, the population size of even 100,000 or less is now considered to be viable in a globalized world. Arguably, the statistical perspective lacks a theoretical justification for any of the population thresholds that the literature considers.

1.3   THE MEASUREMENT DEBATE ABOUT SMALLNESS

But why did we see a surge of small countries as United Nations members? One could assume that proliferation may be connected to a decrease in cost for small countries to become a member of the United Nations. Since the French Revolution, the costs for countries to become independent have increased exponentially with the creation of the nation-states and big empires. Smallness was prone to conquest and was an invitation for aggression. Survival costs, such as administrative and defense costs, arguably, were very high. Alesina argues that this dire expectation about smallness changed with the increase in the degree of hospitableness in the international setting. The degree of hospitableness relates to the trend of global economic integration, which is a salient pattern of the post-war arrangements. As long as there is open international trade available, smallness should become less relevant.[16]

---

[16] See Alesina (2000); and Alesina, Spolaore, and Wacziarg (2000).

However, while there is a proliferation of more small countries as UN members, we also notice that the population size of countries increased during the post Second World War era. Alesina estimated in 2003, the population median at less than 6 million. This estimation fits countries such as Norway, Finland, or Slovakia. Our median estimation in 2020 suggests a population size of nearly 9 million characterizing countries as diverse as Scandinavian countries, some Eastern European countries (e.g., Bulgaria), and Central American countries (e.g., Honduras). Does the median entail a cut-off point those countries below the median can be considered as small and countries above the median are large? A cut-off point without any theoretical justification does not provide many insights into the differences in the economic and social potentials anchored in size. The cut-off point only illustrates the distribution of countries by population size.

But can we just ignore the relevance of country size? According to Kuznets, the answer is negative. He asserts that country size explains and predicts country behavior. A small country characterized by size is more likely to engage in trade than a vast country. Trade is important because a small country reveals a tendency to concentrate production on a limited number of activities. Therefore, "Foreign trade is of greater weight in the economic activity of small nations than in that of larger units" (1960, p. 18). But why then is international trade relevant for country size? The benefits of country size seem intimately related to the market size. As we discuss later, market size connects with growth, productivity, and specialization. To overcome smallness and to grow, openness becomes crucial. According to Alesina and Spolaore, "Size and prosperity do not seem to go hand in hand" (2003, p. 81). Therefore, and clearly, assessing, and understanding the relationship between smallness, openness, growth, prosperity, and development is imperative.

Yet, the literature assessing size and prosperity seems bogged down in an endless debate about the nature, direction, and magnitude of this relationship. From the ongoing debate, it is still not clear how this relationship impacts the opportunities for small countries to prosper. The ongoing debate partially struggles with the pessimistic views manifested in mainstream economic thought that smallness is detrimental to economic growth. Small countries cannot grow because they are too small, lacking the required resources to provide for themselves. The costs of producing anything by these countries are too exorbitant because they lack scale in providing public goods and services and market goods. For example, there are high costs of infrastructural construction and utilization per capita (e.g., roads, and telecommunication) and high per-unit costs of training the specialized workforce.[17] The high costs are simply because few taxpayers pay for these goods and services. Consider, for example, a population of 100,000 compared to 17 million paying for defense, a monetary and financial system, a judicial system, infrastructures for communication, police, and crime prevention, public health care, and schools. The per capita costs of these services decline with the number of taxpayers.

Since country size seems relevant in determining the probability for small countries to grow and prosper, the literature dissecting smallness expends attention to operationalize smallness grounded on resource-based assumptions. Different measures have been used to define "smallness," including population size, land area, and GDP. Of the three measurements, population size is the most popular measure, because the population and land area reflect the size of an economy's factor endowments. The population remains the most widely used resource with which to characterize smallness.

The literature advanced three reasons for the desirability of the population metric. First, population numbers are readily available; second, the cut-off point between small and large countries can be precise, and third, population size correlates with other quantifiable metrics such as the size of the economy. While this view recognizes that smallness is multi-dimensional, population size is a reliable, useful, and practical metric. The literature identifies several population sizes used over time. The population cut-off points range from 10 million, as suggested by Kuznets in 1960, to the 5 million of Demas in 1965, 1.5 million of the Commonwealth Secretariat and the World Bank, and 1 million of Hein in the 1980s, to 500,000 and 250,000 in the 2000s.

The endowment hypothesis has a strong tradition in comparative theory, and it holds that the environment influences land quality, labor, and production technologies—all of which shape economic development. This perspective assumes that a larger land area has the potential for more abundant natural resources. Kuznets asserts "… a large area will have a much greater variety than a small area-of minerals, of climate, of topography, of mixture of land and water …" (1960, p. 17). The implication is, generally speaking, that natural resources are likely to be limited and

---

[17] See, for example, Easterly and Kraay (2000); and Gal (2003).

undiversified in a small country. The geography framework asserts that geography may prompt income because naturally endowed resources and their quality largely depend on geography. However, empirical evidence suggests that the path towards prosperity is not the existence of abundant resources, but how a country uses those resources to fuel prosperity. Small islands, because of their limited geography, can escape the resource curse. This resource curse bequeaths those countries with abundant natural resources with rent-seeking and rent-distributing institutions that hinder growth and economic performance.

The other characterization of smallness is gross domestic product (GDP). GDP is a frequently used measure for market size because it reveals all goods and services that a domestic market (businesses and government) produces and provides each year. GDP is closely related to the incomes generated in a domestic market.[18] These incomes refer to profits and wages; when the economy grows, measured by the GDP, it means that incomes have increased either through higher profits, savings, and wages, lower unemployment, or both. An essential source of growth is the size of a population. However, GDP as a measurement of the economy should be used with caution. For example, GDP does not capture all the information about how well we are doing. It does not reveal the degradation of the environment, traffic congestion, income distribution, and its social impact; neither does it say how well off we are as individuals and country. Nevertheless, because GDP is highly correlated with population size, population size has played a dominant role in the development viability debate.

1.4   A BASIC PUZZLE

The literature suggests that small country size impedes economic growth. However, the empirical evidence reveals that small islands do not necessarily experience a significant disadvantage for their economic development, although growth trajectories seem uneven and erratic. As Easterly and Kraay put it in their article *Small States, Small Problems*? "If small size is a disadvantage, then these states must suffer with a vengeance." The discrepancy in economic growth and prosperity among various small island countries has become a tantalizing research topic for scholars. Overall,

---

[18] The connection between GDP growth and wage/unemployment growth does not always exist. In recent years, economies have seen both wage and job stagnation (jobless growth or jobless recovery).

size has played a dominant role in the literature regarding the economic, political, and social development of countries.

Large countries engender scale, higher productivity, lower unit production costs, and safety compared to smaller countries. Also, some commentators regarded large countries better suited to handle the diversity of preferences and effects of fractionalization.[19] For example, Madison in the *Federalist* (X), expressed,

> *The smaller the society, the fewer probably will be the distinct parties and interests composing it; the fewer the distinct parties and interests, the more frequently will a majority be found of the same party; and the smaller the number of individuals composing a majority, and the smaller the compass within which they are placed, the more easily will they concert and execute their plans of oppression.*

Therefore, Madison advocates for larger units and against smallness to protect people from oppression. However, the belief that larger units may stem oppression is not widely shared among commentators. The increase in heterogeneity of preferences seems more likely to lead to oppression. Alesina argues that if larger units were only beneficial, than the world would consist of large countries. If largeness implies lower production costs per capita, a buffer against foreign aggression, more significant domestic markets, bigger social nets, and internalization of externalities, then why is it that the world does not consist only of large countries? It is because of the extant trade-off that exists between heterogeneity of preferences and size, according to Alesina. Largeness consists of a myriad of preferences, culture, and attitudes of the population. The administration costs increase with the divergence in preferences to the point that when the equilibrium between costs and benefits is broken down, largeness suffers.

The implication is that countries become smaller because the costs induced by the heterogeneity of preferences and the benefits of largeness interrupt their equilibrium. In Alesina's view, this is the reason for the borders revealed in the Balkans. Alternatively, smaller countries seem more vulnerable, suffering from limited possibilities, and manifesting danger due to foreign aggression.[20] However, others, like Easterly and Kraay, considered that size is not relevant in determining the economic

---

[19] See, for example, David Hume and James Madison, who were proponents of this view.
[20] See, for example, Harden (1985).

performance of countries. These authors assert that small countries only face 'small problems.' These opposite views still pit scholars against each other.

Throughout history, one can identify countries with a very diverse area, from China and India, as vast countries in terms of area to Nauru and Tuvalu, which are miniscule islands consisting of about 10 square miles each. This diversity of areas in the world poses the question of whether size matters for economic success; if so, how? The primary consideration centers on what the relationship is between country size and economic performance. This study defines economic performance as the level of income per capita.[21] This level of income per capita is an indication of the ability of a country to provide for the required goods and services to its residents. The definition also embraces the notions of efficiency (allocation of resources) and effectiveness (the achievement of welfare). Typically, studies investigating the impact of country size on economic performance compared income levels and growth rates among small and large countries and the significant challenges they face.[22]

1.5   POPULATION AND SMALLNESS

The question regarding the relationship between country size and economic performance is relevant and meaningful for several reasons. The previous discussion illustrates a mixed picture, a changing meaning of smallness regarding its relationship with growth and prosperity. While mainstream economics (neoclassical paradigm) identifies the relationship between smallness and growth as a non-starter, empirical evidence suggests the

---

[21] Performance is an economic concept directed at results over time. Typically, development economics embraces economic growth and quality of life as the desired results of development. In keeping with Sen's views, we can view economic development in terms of the means and extent by which a country's capacity to drive and command the way its society achieves the kind of success that would be of benefit to the constituency that comprises a competitive market state: namely citizens, business potentials, and thus the community at large. The capacity to achieve success on a competitive scale involves a collective competency to address unforeseen problems and issues that would naturally arise in the course of development. Of vital significance is the function of human capital in their ability to address economic development along behavioral configurations that incite and sustain the kinds of qualities attributed to competitive progress-namely, understanding that their ongoing and consistent participation in economic development hinges on the role they play in building a cooperative avenue between themselves, business, and community needs.

[22] See, for example, Armstrong and Read (2001); Briguglio (1995); Read (2004); Selwyn (1975); Shand (1980); and Srinivasan (1986).

opposite. For example, James Meade, a 1977 Nobel Memorial Prize winner in economic sciences, asserted that Mauritius, at the threshold of its independence, was a strong candidate headed towards economic failure. Insofar as small islands, such as Mauritius, do not reflect the necessary wealth of resources to support economic growth, and insofar as its own market was prohibitive, it would naturally be predestined for economic failures. Yet, over time, Mauritius' development was more than adequately successful.[23] Other studies including Easterly and Kraay, Armstrong, and Read and ECLAC[24] suggested that smallness and growth have a positive relationship.

We must measure what is relevant. Population size may not be in sync with how countries experience the impact of smallness on their citizens' lives. This premise prompted an analysis of the relationship between population and economic performance. The sample of 54 islands around the globe reveals significant variance in the population size of the countries ranging from Japan with a population of 126.5 million to Montserrat with nearly 5,000.[25] The analysis centered on 13 variables defining economic performance, i.e., real GDP (constant 2015 prices), GDP per capita (constant 2015 prices), inflation (average consumer prices), unemployment rate (% of the total labor force), population, gross government debt (% of GDP), current account balance (% of GDP), international tourism arrivals, international tourism arrivals divided by population, tourism receipts in US dollars, tourism receipts as the percent of exports, the ratio of tourism receipts to GDP, and contribution of travel and tourism to GDP.

The analysis used a Single-Equation Cointegration test and the Engle-Granger method. A p-value lower than 10% estimated the cointegration threshold of four population groups, Finally, an interquartile range (IQR) test estimated the ratios between the numbers of no-cointegrated to cointegrated combinations. The results in Table 1.2 reveal that population is not a critical factor in determining economic performance. The result suggests that the sample may be homogenous as an island group but is rather heterogeneous based on economic performance. If not population, then what is the critical criterion that distinguishes smallness?

---

[23] See James Meade et al. (1961). Meade's reading of the economic possibilities turned out not to be correct as Mauritius, over time, resulted in a success story in Africa. See on this matter, Subramanian and Roy (2003); and Seetanah et al. (2019).

[24] See, for example, ECLAC (2001).

[25] The analysis used multiple sources such as UNSTATS, ECCB Annual Economic and Financial Review, World Bank, CIA World Fact book, IMF, WTTC, UN World Population Prospects.

TABLE 1.2 Ratios between Cointegrated and Non-Cointegrated Countries by Population Segments

Indicator	Population				Ranges for Outliers	
	≤ 1 million	≤ 1.5 million	≤ 3 million	≤ 5 million	High	Low
Real GDP (2015 = 100)	5.4	5.5	5.4	5.5	5.7	5.2
Real GDP per capita (2015 = 100)	5.6	5.5	5.5	5.6	5.7	5.4
Inflation (average consumer prices)	4.0	3.8	3.6	3.7	4.3	3.3
Unemployment rate	14.5	15.3	15.2	17.6	18.8	12.6
Population	14.0	14.0	12.7	12.2	16.4	10.0
Government gross debt (% of GDP)	7.8	7.9	7.9	7.8	7.9	7.7
Current account balance	4.0	4.6	4.7	4.8	5.4	3.7
International tourism arrivals	6.1	5.3	5.4	5.8	6.8	4.5
International tourism arrivals divided by population	5.3	5.0	5.1	5.4	5.8	4.7
International tourism receipts (percentage of total exports)	10.2	8.2	8.6	9.0	11.5	6.6
International tourism receipts (US$)	15.1	15.5	16.0	15.5	16.5	14.6
Ratio of international tourism receipts to GDP	9.6	9.2	9.4	9.7	10.3	8.7
Contribution of travel and tourism to GDP	6.5	4.9	4.9	5.1	7.1	3.6

1.6   TOWARD AN ALTERNATIVE FRAMEWORK: THE SMOP

We can distil from the previous discussion two conclusions. First, small-ness evokes a distinct category of countries. There is a long provenance to support this conclusion, which garnered more endorsement in the WWII postwar era. The premise of this endorsement is that small islands could not emulate the developmental trajectory of large countries. It appears logical that to follow the trajectory of large countries, a small island should be endowed with sufficient resources-resources that reveal large quantities and quality of factors of production. It is almost unfathomable to grasp this thought if one thinks about the small area, the small labor pool, and narrowly qualified labor supply so pervasive in small islands. There is abundant literature that mentions that small islands, in general, are resource-poor environments. Commentators suggest that small islands are unique with distinguishing characteristics. These characteristics suggest substantial disadvantages that small size imposes on growth processes, as indicated in the previous chapter.

Multifarious complexities can bill small countries as possessing difficul-ties too convoluted to keep pace with the more developed countries. Among these are their limited resources with which to build their performance, particularly in a sustained manner, thus rendering their efforts too close to futile. Further, their degree of openness may unfairly label them as takers and beggars within the global market, thus compounding positive perfor-mance markers when compared to developed countries. That complexity is made, especially officious of their growth performance. Also, as 'income penalties' accrue to businesses that situate themselves in small islands, nega-tive economic impacts arise that jeopardize such liaisons.

Continuing theories abound that for a country to grow, to progress, and to develop, that country needs to have abundant resources, particularly labor supply. One of the theories that center on explaining how developing countries can grow is the Lewis industrialization model.[26] This model claims that industrialization is the engine of economic growth. The model alludes to two conditions for industrialization to occur, namely a large agricultural sector and a large labor force. Small islands cannot meet these two conditions. The immense labor surplus that Lewis envisages to propel a country from stagnation to self-sustaining growth is sorely absent in a

[26] See Lewis (1955).

small island. Therefore, as conferred by literature, agriculture's role in the structural transition necessary for industrialization is quashed for small islands. The reason is that agriculture is a low productivity sector, and the labor pool is too small to compete with larger countries in labor-intensive industries. While in the time of Lewis, it may have been construed otherwise; today, the small labor pool and the subsistent nature of the small island's agricultural sector can be construed as an advantage. This advantage standpoint has been eloquently unearthed by Armstrong and Read[27] (p. 104),

> *This scarcity of labor can be regarded as a distinct source of advantage ...since traditional agriculture is generally viewed as hindering growth because of its low productivity and inferior technology. It also rules out labor-intensive industrialization, however, since small states cannot hope to compete with larger industrializing countries in low-skilled labor-intensive export sectors.*

The corollary of Armstrong and Read's suggestion is that small countries do not follow the growth stages of large countries and that they are a unique set of countries. Small islands are, by their nature, impeded in generating a critical mass of domestic activity that is sufficient to achieve a minimum scale necessary for efficient output. Armstrong and Read qualify the smallness feature as churning "sub-optimal economies." Sub-optimal economies distort prices to the detriment of the welfare of people. The reason is that small-scale affects all three dimensions of efficiency, namely allocative, productive, and dynamic. The lack of allocative efficiency induces distorted factor costs and affects relative utility. The distortion in costs means that output is not produced at its lowest costs (productive efficiency), which dents the incentive ecosystem affecting productivity and innovation (dynamic efficiency). Gal considers these three scale shortcomings as the key characteristics of small economies, converting these small economies as a distinguished group of countries.[28]

The second conclusion is that smallness conjures a notion of vulnerability. That is, small island countries are given to external shocks that likely cause extreme survival difficulties for their communities, businesses, and future economic prospects. This likely exposure to shocks increases

---

[27] See Armstrong and Read (2003).
[28] See, Gal (2003). Chapters one and two of his book provides an extensive discussion about the economic shortcoming of small economies.

their vulnerability and curtails their existing and prospective long-run growth. Vulnerability is connected to exposure to natural disasters, social, and economic conditions, and uncertainty regarding climate change. The concern is that vulnerability compromises the welfare of small countries. However, is this notion of smallness useful and adequate? Our answer is no! We take issue with this characterization of smallness. I am not rejecting vulnerability as an essential aspect of smallness, but I attempt to draw a nuanced picture of smallness based on the choice made to grow and prosper. Considering smallness as vulnerability may lead to paralysis, inaction, and apathy. It is like looking at a glass half empty instead of half full. If you are vulnerable and the reasons for being vulnerable are factors beyond your control (such as geography), then there is no sense in trying to overcome smallness. Indeed, would this then not be a death knell to the small island population?

Geography will always doom smallness to vulnerability unless a small country is lucky to possess some critical resources (e.g., oil, and gas). If geography is so unforgiving, then why is it that there are small countries that have economically and socially performed better than others? That is, if smallness dictates similar characteristics among many small islands (resource issues, etc.), then it might be assumed that these deleterious traits would yield similar economic development statuses among countries. Yet, this is not true of the prosperous Caribbean countries where they can boast of their affluence, resources, and prospects.[29] For example, in the Caribbean region, one of the wealthiest countries in the world in per capita income, The Bahamas, exists alongside one of the poorest, Haiti.

Real income per capita ranges from, for example, US$ 868 in Haiti to US$ 32,997 in the Bahamas at a ratio of 1 to 38. GDP per capita in the Bahamas is also more than six times larger than Jamaica (US$ 5,268). The Bahamian population is 39 times and 10.5 times smaller than Haiti and Jamaica, respectively.[30] A similar situation reveals itself in the case of the small island in the European Union (EU) uncovering several factors impacting their growth performance.[31] Moreover, small size cannot empirically or undeniably dictate failure or even overly limit the economics of growth. Indeed, while issues of external shocks and vulnerability may

---

[29] See, for example, Croes (2005); Wint (2002); Griffith (2002); Escaith (2001); Bernal, Bryan, and Fauriol (2001); Freckleton (2000); and Modeste (1995).
[30] Haiti is technically not an island but unites with the Dominican Republic, the Island of Hispaniola.
[31] See, for example, https://www3.gobiernodecanarias.org/aciisi/ris3/documentos/otros/rup/15-estudio-de-los-factores-de-crecimiento-en-las-rup-resumen-espanol/file.

loom, it cannot be assumed that size or looming threat disqualifies the potential for success or success itself. From this perspective, geography is just geography.

Another reason for contesting the vulnerability notion is that this notion lacks theoretical underpinning. Vulnerability cannot explain or predict growth and prosperity in small countries. The most vulnerability can explain are the challenges lying ahead as induced by smallness. In other words, vulnerability is more an effect of smallness than its cause. The notion of vulnerability also is static because it just refers to the position of a small country, but not how this small country moves from one position to the next or back. This observation means that the notion of vulnerability lacks the dynamic aspects of smallness. Vulnerability neglects the opportunity to make geography fungible through the application of trade. Also, there is empirical evidence that relates vulnerability positively with economic growth. The reason is the dominant positive role of openness in economic growth.[32]

Smallness can become significant when we consider the economic and social activity of international trade. If smallness of the domestic market is the problem due to the extent of the small size of demand, then opening up by integrating international trade in domestic, economic, and social activities could overcome the problem of smallness. For small countries to harvest potentially lucrative means to obtain economic growth and development goals, they have had to look to trade. The impetus of trade toward those goals has impacted growth that has surpassed the economic growth beyond other countries. Indeed, empirical evidence reveals small countries' growth rate has transcended larger countries' rates by over three times.[33]

Openness, which is the hallmark of small countries, enables the integration of foreign markets into the domestic markets of small countries. This openness and its commensurate integration in domestic economic and social activities are intertwined with the development potentials of

---

[32] For example, Briguglio et al. (2009) found that the GDP per capita was negatively related to vulnerability while being positively related to resilience. However, they found that GDP per capita was more responsive to resilience when compared to vulnerability. They conclude that policies were more important than structural economic characteristics of smallness. Pereira (2018) found, for example, that while Aruba was the most vulnerable island in the Caribbean, Aruba was also the most resilient. See also Read (2010).

[33] See, for example, Helpman (2004) who assessed the effects of trade on economic growth comparing Mali and Seychelles.

smallness. Thus, viewing the potential of small islands to achieve economic growth through the framework of SMOP recognizes that the relationship between *SMallness* and *OPenness* (SMOP) is a valuable operation with which small islands might foster their upward mobility.

Clearly, openness consummates itself in the much desired and needed goal of growth. A small country's resources, such as natural and human resources, are fixed in the medium term and very challenging to change in the long run. The way to change these resources is through international trade. International trade increases the domestic market facilitating more economic activities beyond the rigidity of the small domestic market. Moreover, international trade increases demand for local goods and services while providing for the necessary imports, and in so doing, can trigger growth, and importantly, continue to effectuate future growth.

Growth, over time, is a marketing objective that reveals the viability of a nation to secure sustainability. Openness to foreign markets as a strategy for small countries has resulted in positive growth over time for tourism dependent countries such as Aruba, the Maldives, Malta, Barbados, and the Canary Islands. Krueger (1985) has hypothesized that international trade in accordance with the utilization and levels of openness can alleviate the constraint issues that plague small island markets. Further, the international trading venue dictates a competition for goods and services that requires domestic product quality be competitive with the products of larger countries. Thus, progressive, or greater degrees of openness can lead to positive levels of economic output given that the degree of openness is defined in terms of the relationship between the sums of imports with exports to the GDP.

International trade, then, imposes standards through competition that lead quantity and quality of production improvements. The theoretical underpinning for the integration of international trade in the internal production process of small countries is the division of labor of Adam Smith. Smith's division of labor theory entails economic exchange (trading) as a means to enhance efficiency and productivity. This exchange is driven by choice-driven specialization. Table 1.3 illustrates the degree of openness induced by smallness. Perusing suggests that there may be a connection between smallness and openness (SMOP). The outcome of an ordinary correlation test in Table 1.3 indicates that smallness correlates with openness (correlation = −0.24396; p = 0.0783).

TABLE 1.3 Average Openness of Countries 2010–2018

Country	Average Openness	Country	Average Openness
Anguilla	118.65	Malta	297.8
Antigua and Barbuda	99.59	Marshall Islands	139.74
Aruba	146.33	Mauritius	107.12
The Bahamas	80.29	Micronesia, Fed. Sts.	105.67
Bahrain	159.51	Montserrat	109.51
Barbados	87.49	Nauru	88.33
Bermuda	78.37	New Caledonia	54.22
The British Virgin Islands	196.31	New Zealand	55.92
Cabo Verde	103.28	Palau	136.48
Comoros	39.67	Philippines	66.88
Cook Islands	140.99	Puerto Rico	106.73
Cuba	36.55	Samoa	81.14
Curacao	148.64	Sao Tome and Principe	61.71
Cyprus	128.27	Seychelles	190.84
Dominica	91.51	Singapore	347.05
Fiji	113.74	Sint Maarten	217.16
French Polynesia	50.51	Solomon Islands	113.81
Grenada	91.3	St. Kitts and Nevis	98.12
Iceland	95.7	St. Lucia	86.63

TABLE 1.3 (*Continued*)

Country	Average Openness	Country	Average Openness
Jamaica	82.22	St. Vincent and Grenadines	89.18
Japan	33.31	Tonga	83.72
Kiribati	103.9	Trinidad and Tobago	86.37
Macao SAR, China	118.01	Turks and Caicos Islands	137.34
Madagascar	59.47	Tuvalu	126.2
Maldives	153.14	Vanuatu	101.37

Note: Openness is estimated by the sum of exports and imports divided by GDP.

What we know is that smallness consists of openness as a defining component. In the next chapter we discuss the second defining component, which is specialization. Chapter 2 articulates how Smith's division of labor links specialization to overcoming smallness.

KEYWORDS

- gross domestic product
- interquartile range
- minimum efficient scale
- small island developing states
- sociological intimacy framework
- subnational island jurisdictions

CHAPTER 2

THE TOURISM SPECIALIZATION HYPOTHESIS

If one follows the opinions of some previous chapter commentators, an affirmed impression emerges that small islands are doomed to fail. The economic performance to which these islands aspire would be subpar, and they would be hamstrung to meagerly provide for their small population, much less build a sustainable economy. However, is this affirmation resolute? Does it portend the future of small island destinations? Is there no feasible recognition of their potential to grow or of their people to project the kind of intrepid, enterprising spirit granted to people of larger venues? Are small islands relegated only to accept any low hanging fruit that they can grab and hold onto? Is there a way out of such a conundrum-of the misfortunes and afflictions imposed by small size? Must their scale problems prohibit their economic welfare?

One of the roadmaps to overcome scale problems was pointed out by Simon Kuznets (1901–1985). He suggested that trade could compensate for small size because trade enlarges the demand possibility curve for these small islands. If one can reify Kuznets's suggestion, it still begs the question of whether small islands can trade in all kinds of goods. Arguably, small islands' economic development history suggests that small islands cannot trade in all goods precisely because they have factor constraints, such as limited land and labor supply. Thus, scale constraints are an imposing issue impacting small island success. By necessity, then, small island destinations require rigorous consideration for a beneficial means with which to overcome such constraints.

This chapter examines the role of specialization as a channel to overcome scale constraints. In particular, it analyzes the exogenous and endogenous comparative advantage theories about the conceptual foundations of tourism specialization (TS).

2.1 DEFINING SPECIALIZATION

In the previous chapter, we characterize smallness with openness. Smallness leads to openness, and openness leads to specialization. Smallness induces insufficiency and immobility of factors, the economy cannot produce everything that is needed and required by the population, and people do not have all the skills to produce what is needed. Thus, a definition was articulated for smallness based on the logic of that insufficiency and immobility. In other words, production, and consumption are not in sync. Therefore, smallness, by definition, propels openness. However, openness compels specialization in order to engage in foreign trade.

Moreover, how can we define specialization? Where does specialization as a concept come from, and how do we measure specialization? Definitions are important because we cannot clearly explain the meanings and assumptions related to concepts without them. We can also not measure what is relevant and what counts toward expediting the vital benefits of specialization. To advance our understanding of rigorous research and practice, we must develop a precise definition of specialization. Better understanding enables communication among researchers, facilitates the development, comparison, and evaluation of outcomes, and shares knowledge.

Specialization has a long provenance and has fascinated policymakers and commentators because specialization references a place identity, technological, knowledge, or cultural hub. Cities and metropolitan areas in large countries are usually described in terms of their iconic activities, such as high-tech, finance, logistics, services, or labor-intensive manufacturing. Planners, consultants, and business leaders are interested in understanding what causes economic activity patterns in a city or region. This fascination, combined with the urge to understand specialization and its implications on industry welfare, stems even as far back as Plato, who examined these implications on the market and money. Some other classics, such as Xenophon, even considered the link between specialization and urban development. Specialization also caught the attention in the debate about competition among European cities. Fairs and exhibition activities seem an old urban development strategy in Europe that underscores European cities' specialization patterns. Overall, specialization affected the economic size and the international image and reputation of European cities.[1]

---

[1] The study of Cuadrado-Roura and Rubalcaba-Bermejo (1998) provides an excellent account of specialization patterns in European cities.

Yet, what about the concept of specialization? We argue that the concept of specialization entails two aspects. One aspect frames and relates the components constituting specialization, and the other aspect involves a strategy to overcome smallness and engage with openness. The first component references the process, and we characterize that as *a specialization process*. The second component involves the purpose of entailing the specialization process. We call this process the *specialization goal*. The specialization process includes a choice made to broaden the market's extent and the scope of the division of labor to increase productivity and trigger increasing returns and prosperity. Thus, the specialization process encompasses reshaping a country's productive sector of a country to meet the export demand to pursue the specialization goal: economic development, prosperity, and well-being.

The motivation to specialize is to reorganize production, induced by smallness, to pursue new, sufficient demand through trade. This specialization entails involving the use of surplus productive capacity. The choice of specialization involves an inelastic domestic demand for the exported item and the use of the surplus capacity, geared to produce the exportable item. This choice is almost costless because it anchors on the surplus production capacity. It pays for the required imports and the expansion of domestic economic activity to satisfy the domestic population 's needs.

Of course, there is a learning curve involved in identifying and working the product's methods and courses to be exported. Moreover, finding this product is easier said than done because we face a problematic identification and learning process. The domestic population should know and understand what they need, find the other population that can barter or sell what they need, and then trust, indefatigably, that the other population will deliver what they need according to a predetermined agreement. This agreement entails the quantity and quality of the product and the delivery time and timing. The domestic population should be able to barter or to pay for this agreement. The exchange can take place when the domestic population can produce something valuable to the other population group.

Arguably, this process presupposes the self-interest of both parties wanting to engage in product exchange. Party self-interest is at stake because the parties would be expending their efforts, time, and resources. These efforts and resources must respond to knowing and understanding what, why, and how to engage in a production activity that satiates the consumption valuable to both participants-including how costs and bene-fits would be calculated and dispensed to both parties. This calculation presumes a rational behavior that could result in social welfare benefits. Both Adam Smith, through his "...benevolence of the butcher..."[2] and Marshal's reference to the "strongest forces" advocate for each individual or party to enjoy the full scope of that which secures his own interests. Thus, administering to his self-interests merged with that of others eventu-ally infiltrates society's welfare.

Moreover, self-interest deploys the identification of something valuable to the other group. This identification requires knowing and understanding the other party's demand, tastes, preferences, and the fundamental motiva-tion to barter. The motivation is dictated according to their preference for a variety of goods and services. However, it is one thing to analyze a party's motivation; it is another thing to satisfy the preferences that piloted that motivation. The satisfaction of those preferences involves skills and tech-nology of how and when to produce the quantity and the quality of prod-ucts and services that are valuable to the market. The skills and dexterity entail quick and cost-effective learning of those tastes and preferences in the market. Quick learning may provide opportunities to cross-pollinate with other economic sectors, thereby expanding economic activities and enhancing economic growth and development. Thus, production skills, knowing the market, and the effective distribution of a given product is involved.

The learning process necessitates the diffusion of technology and entrepreneurial scholarship to improve products and services' quality. Constant competitive exchange forces the improvement of standards in terms of quality and the expectations not only to construct quality but also to receive it. However, the outcome of the considerable portion of time and efforts invested by the population (individuals) in these economic activi-ties is neither predictable nor inevitable. While the intent to extract the exchange's mutual value is clear, the activity may not yield the expected outcome. The probability outcome of the activities hinges on trust: trust to complete the exchange process in a manner both parties ascribed to per the initial agreement. Trust can become an amplifying resource through

---

[2] The complete phrase in Adam Smith's *The Wealth of Nations* is "It is not from the benevolence of the butcher, the brewer, or the baker that we expect our dinner, but from their regard to their own interest" (Book 1, Chapter 2). This phrase suggests that self-interest can lead more effectively to the collective good.

the process of "learning-by-doing" in that its availability is a function of its use.

However, enhancing trust is a delicate balance in the context of simultaneous competition and collaboration among agents. If this balance is not optimal, the capacity to "learning-by-doing" may be eroded which makes its mirror image also true: trust not used may become depleted. Thus, any exchange value deviates over time using or disusing skills, trust, and ability. Albert Hirschman uttered this reality when he asserted "human abilities and skills are valuable economic resources, most of them respond positively to practice, in a learning-by-doing manner, and negatively to non-practice." [3]

2.2   RICARDO'S COMPARATIVE ADVANTAGE

I discussed in the previous section the attributes that form the concept of specialization. Now is the time to discern the ontology of these attributes as they constitute specialization. The tourism literature centers its analysis on tourism flows on Ricardo's comparative advantage. David Ricardo anchored his notion of comparative advantage in his *On the Principles of Political Economy and Taxation*. Ricardo articulated the idea in 1817 that countries engage in international trade even when one country is more efficient at producing every good. He used the England-Portugal example in producing cloth and wine to illustrate how production costs differences trigger trade between countries. This example reveals the comparative advantage theory of David Ricardo. A country or an individual specializes in an activity in which it is relatively more labor efficient. Applying this specialization rule fosters welfare-enhancing effects revealed in mutual gains for everyone. The underlying condition for the existence of trade is different from comparative production costs. Ricardo's notion of specialization departs from natural differences amongst economic agents and assumes that production factors are entirely used. The purpose of Ricardo's trade is the reallocation of resources and depicts a movement along a production possibility curve.

His approach presumes that comparative advantage or advantages lie outside the economic activity or the individual (tastes, technology,

---

[3] See Hirschman (1984), (p. 94). The person who coined the phrase learning-by-doing is Kenneth Arrow (1962).

or endowments). For example, an island may be endowed with beaches, betting its comparative advantage on that endowment. The beach cannot be changed by any endogenous factor, such as learning or technology; only exogenous factors shape comparative advantage. Exogenous factors are as "those resource endowments which cannot be changed by any endogenous factor in the correspondent country economic system."[4] The existence of a beach is outside the destination's control and may become the primary motivator for tourists to travel to that particular island destination. Besides natural factors, such as scenery, mild climate, and landscapes, heritage factors are also exogenous. These factors include history, music, and special events. Within this latter category of exogenous factors are the World Cultural and Natural Heritage Sites by UNESCO.

Ricardo's premise of comparative advantage rests on the notion that the two economies are closed at the time of production, while open at the exchange point where the economy settles into equilibrium. The Ricardo logic locates trade's origins in differences in natural capacities to produce and in the prevailing patterns of demand to create value. The production costs may differ due to different endowments of factors of production.[5] The advantage is acquired by natural differences, which reveal production cost differences. The implication is that countries should specialize in activities of lower production costs, which depend mainly on abundant factors of production. Trade is welfare enhancing according to this perspective because trade pushes each country's consumption possibilities beyond domestic production. Thus, the two countries progress economically if they can realize their comparative advantage by reallocating resources and trading with each other. They can grow more than others due to supply-side factors, such as the elasticity of demand.

Ricardo's comparative advantage theory warrants three critiques. First, the theory can only explain one-time benefits due to trade, which means that it cannot enlighten dynamic economic activities, triggering either gains or losses.[6] This means that his theory is a static comparative model. Dynamic developments, such as economic growth, do not form part of Ricardo's model. Second, the foundation of the theory is supply-based. The ability to be more cost-efficient to exploit the existing natural resource

---

[4] See Hong (2008), p. 54.

[5] The different production costs are the hallmark of the Heckscher-Ohlin model. This model centers on the assumption that specialization results from the abundance and cheap factors of production.

[6] For a critical discussion about Ricardo's comparative advantage, see, for example, Schumacher (2012).

base, for example, is not a guarantee for growth and prosperity. Otherwise, all developing countries whose positive trade balance owes to their export of natural resources would be in a better economic position. The theory does not include room for market-creating opportunities. Moreover, third, the source of Ricardo's comparative advantage is exogenously given, namely productivity differences or different endowments. In this view, differences between countries remain constant once trade starts.[7] The standard comparative advantage approach may be inadequate to address the fundamental problems of small islands.

Only a handful of tourism studies embraced Ricardo's comparative advantage framework.[8] These studies assert that supply-side approaches may better explain tourism flows compared to demand studies. The interlocking of comparative advantage stems from technology anchored in labor efficiency and available endowments such as abundant available resources (e.g., beaches, mountains, and labor). Opponents to this view assert that these studies only assess tourism flows and patterns as a one-time event. This static examination of tourism patterns cannot explain the dynamic positioning of destinations over time. The embeddedness of comparative advantage in endowments and price elasticities seems to counter the empirical tourism literature. That literature suggests that the most critical determinant of tourism demand and growth opportunities include income elasticities. Perhaps it is not so essential to query whether a country engages in specialization and trade; instead, it seems more important to examine specifically what trade the countries specializes in. Adam Smith might provide some insightful answers to this query based on his division of labor framework.

2.3   ADAM SMITH DYNAMIC COMPARATIVE ADVANTAGE

Adam Smith, in his *Wealth of Nations*, ensconced specialization as the result of the division of labor described in pin factories. Changing the

---

[7] That specialization is a suitable course for nations that have been deprecated by the Singer-Prebisch thesis, along with Keynesian economics, in that terms of trade can be fickle-changing over time. Also criticized has been the comparative advantage theory, characterized as inert-not allowing for economic development influences over the long-term. Warranting this claim is that investments and subsequent developments can provide advantages that impact economic development in the long-term.

[8] For a view on the application of Ricardo's comparative advantage, see, for example, Hassan (2000); Sahli and Nowak (2007); and Zhang and Jensen (2007). The study of Nowak, Petit, and Sahli (2010) reveal the opposite result.

organization of work through the division of labor allows for an increase in specialization and hence productivity. Smith suggests a logic cemented on the premise that trading or exchanging is a more efficient means of production. Because there are mutual gains from trading, it is logical to exchange. In this context, it is impossible to know in advance what skill set a person possesses because Smith does not consider that skill set differences among people are as impactful as specialization. It is merely the advantages of the specialization circumstance that generate the opportunity to exchange. Yet each of us can produce more economic value by doing one thing than doing a little of everything ourselves. (Kind of a jack of all trades, master of none perspective).

Specialization increases the variety of goods and services. The variety of goods and services is low in less developed and autarkical economies and is zero, where everyone self-provides everything needed. The implication is that specialization happens by choice and not due to natural differences. However, how does specialization facilitate growth or "wealth of nations"?

The degree of specialization, in Smith's view, depends on the extent of the market. In his book 1, Chapter 3, he mentions:

> *"As it is the power of exchanging that gives occasion to the division of labor, so the extent of this division must always be limited by the extent of that power, or, in other words, by the extent of the market. When the market is so small, no person can have any encouragement to dedicate himself entirely to one employment, for want of the power to exchange all that surplus part of the produce of his labor, which is over and above his own consumption, for such parts of the produce of other men's labor as he has occasion for" —Adam Smith (1776).*

The quote implies that the "wealth of nations" increases directly with the market network's size. The network's size relies on the market agent (a person) that exchanging economic products with other agents creates more economic output and activities than engaging in an autarkical activity. The exchanging process entails a continuous improvement of skills, reconfiguration of factors, and increasing returns to use the surplus resources to meet export demand. The incentive to exchange is the preference for a variety of goods. Therefore, the exchange is the deliberate outcome of a choice to engage with others. Smith narrates this assertion as follows:

"Between whatever places foreign trade is carried on, all of them derive two distinct benefits from it. It carries out that surplus part of the produce of their land and labor for which there is no demand among them and brings back in return for it something else for which there is a demand."

Comparative advantage can only exist when there is a decision about the level to specialize. This choice is the source of endogenous comparative advantage.[9] While the division of labor leads to quantitative and qualitative production improvements, the market network determines the division of labor's scope and scale. The market network is dependent on whether the produced goods or services can be sold. If the market expands, more division of labor is possible, and, in turn, would lead to economic growth and more division of labor. The implication is that specialization begets specialization. In other words:

$$\text{Market} \rightarrow \text{Specialization} \rightarrow \text{Market} \rightarrow \text{Specialization}$$

This virtuous circle from demand to specialization to more demand and more economic growth is the crux of Smith's dynamic vision of comparative advantage. This logic's corollary is that international trade may extend the domestic market because "… of the exchangeable value of the annual produce of the land and labor of the country."[10] The extent of the market seems to anchor on population size and purchasing power. The latter hinges on productivity growth due to labor division and the coordination of transaction costs to trade goods and services.

The market indicates a compromise between a buyer and a seller about something (either good or service). Both components (i.e., buyer, and seller) must be present for a market to exist. They make a compromise on the value of that particular thing that is revealed in efforts, quality, and the intensity of wants and needs. The value of the thing is revealed in the price of what the buyer is prepared to pay. Price is propagated by scarcity and the preferences of economic actors. Smith claims that productivity differences, together with demand preferences, determine society's economic surplus benefits.[11]

---

9 See, for example, Yang (1994); and Schumacher (2012).

[10] See Smith (2005), IV, iii, C.3.

[11] Adam Smith first alluded to the concept of absolute advantage as the basis for international trade in 1776, in The Wealth of Nations: *"If a foreign country can supply us with a commodity cheaper than we can make it, better buy it off them with some part of the produce of our industry employed in a way in which we have some advantage. The general industry of the country, being always in proportion to the capital which employs it, will not thereby be diminished [.] but only left to find out how it can be employed with the greatest advantage."*

The idea that demand is critical to prompt specialization and market extension also resonates with modern economic thinkers. For example, demand became a cornerstone of Keynes *General Theory*, explaining economic cycles prompted by demand expectations. Years later, Linder posited that trade patterns derive from "overlapping demand." According to Linder, countries generally produce goods for their domestic markets and then export the surplus to countries with similar tastes and preferences. Hence, the expectation is that most trade happens between similarly endowed countries.[12] Kaldor also emphasized the relevance of the demand dimension in economic growth and posits that trade performance is the primary determinant of growth.[13]

Thus, trade can allow countries to attain the desired increasing returns to scale because trade expands the domestic market by integrating the international market to create new market opportunities. These new market opportunities include new customers by making products accessible and affordable, enabling many more people to buy and consume. Trade has the advantage of increasing choices and lowering prices through competition. There are also quick learning opportunities provided to local business people when they interact with the buyer. These quick and low-cost learning opportunities may be particularly pervasive in the tourism trade when local businesspeople interact with tourists on-site and learn about their tastes and preferences. This learning opportunity may ensue in product development and linkages with other local economic sectors.[14]

The endogenous comparative advantage seems to provide an exciting growth avenue for small islands, namely via income elasticity. Income elasticity implies that there is a potential for residents of these small islands and tourists from source countries to benefit simultaneously. This simultaneous beneficial linkage stems from expanding opportunities and choices available to residents due to tourism specialization (TS) through income and possessions. At the same time, tourists enjoy memorable experiences that can have a positive impact on their well-being. Therefore, the tourism product can ensue in meaningful welfare benefits for both residents and tourists alike. This mutual benefit is a positive surplus from trade. Differences in income elasticity may explain why some small islands perform better than others. Income elasticity implies that it matters in which

---

[12] For an interesting discussion of Linder's hypothesis, see, for example, Hallak (2010).
[13] I discuss Kaldor's theoretical input in chapter six.
[14] See, for example, Lejárraga and Walkenhorst (2010).

activities a country is specialized. Income elasticity may also expound the quality characteristics of the supply components and factors of a destination. Moreover, finally, income elasticity may inoculate the scale challenges so pervasive in small destinations. Remarkably, the tourism literature paid scant attention to Smith's endogenous comparative analysis.

To wrap up, we can say that specialization resolves two concurrent trade-offs. The first trade-off connects specialization with smallness through transaction costs, which are lowered with increased trust thus facilitating trade. The second trade-off relates specialization with smallness through production costs. The first trade-off is the channel through which specialization extends the market to overcome smallness. The second centers on how specialization reduces the high production costs inherent in smallness. These channels interact with others continuously, either regressing or progressing in their intensity and frequency. The outcome of these interactions determines the level and rate of economic growth, prosperity, and well-being in small countries. The chart below shows how these two concurrent components and patterns relate to each other in dynamic contemporaneous and space dimensions (Figure 2.1).

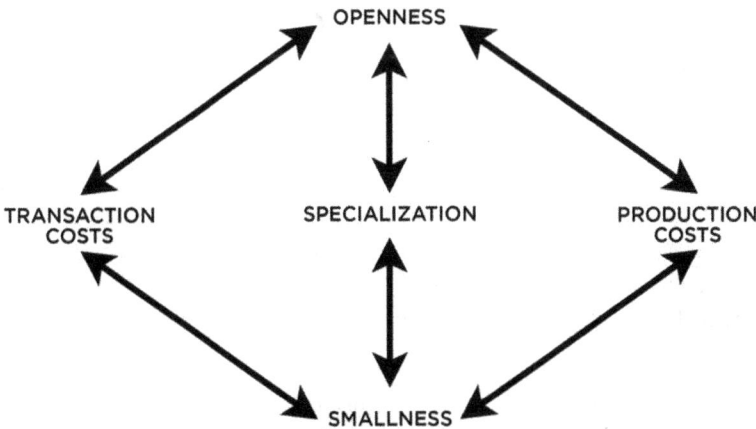

FIGURE 2.1 Two trade-offs illustrating dynamic specialization.

2.4 TOURISM SPECIALIZATION (TS) AS A DYNAMIC PROCESS

Tourism studies seem to assume that TS relies on the comparative advantage premises anchored in trade theory as posited by Ricardo. We already referenced these tourism studies. I take issue with this characterization of TS positing that TS aligns better with the fundamentals of Adam Smith as enshrined in his division of labor theory. I define TS as a collective process-not a one-time event-resource allocation and engagement in the market that creates opportunities to facilitate economic value-added and personal choice (human development). TS entails a combination of demand- and supply-side approaches immersed in demand elasticities influenced by supply constraints.[15]

Thus, TS specifically anchors on the simultaneous involvement of supply and demand features that include the coordination of multiple economic and social agents (i.e., domestic consumers, tourists, producers of non-tradable goods, and producers of tourists' goods and services) to provide a unique attractiveness. This attractiveness features an experience (what); attractiveness also embodies a management process to create this experience (how), which presumes an understanding of the determinants of demand (why). This definition captures the process as the outcome of the alignment of demand and supply through the measurement that entails the ratio of export to the gross domestic product (GDP). This measurement counts because TS entails generating foreign exchange earnings to finance imports and trigger economic growth, favors employment, and increases in income, and TS generates economies of scale and scope. This measurement directly relates to the primary purpose of adjusting smallness issues, i.e., overcoming the domestic market's narrowness!

Studies have shown that TS generates externalities that may trigger structural change by shifting economic activities favoring other macroeconomic sectors.[16] Structural changes are interrelated processes across economic sectors and reference modification in the relative importance of the economic sectors; and these changes shift the economy to more advanced levels. These changes may also affect institutions and societal

---

[15] This perspective explicitly acknowledges the government as an active agent in creating destination value and wealth, contrary to the neoclassical economics perspective. For example, Clancy (2001) illustrates the Mexican government's role in exporting paradise through infrastructure development, loans to the industry, laws, tourism policies, and international tourism marketing. In my *Anatomy of demand in international tourism*, I discussed the Aruban government's active role in everything concerning the tourism industry from 1986 onwards.

[16] See, for example, Marsiglio (2018). Balaguer and Cantavella-Jorda (2002) suggest that tourism financed Spanish structural change by bringing foreign exchange. Hernandez-Martin (2008) asserts that the shift from the agricultural sector to the tourism sector triggers economic growth. McLennan et al. (2012) show how tourism propagates structural shifts across macroeconomic sectors, often replacing traditional agriculture sectors.

values leading some to characterize these transformations as modernization. TS, thus conceived as a social concept, is a way to tackle the economic, organizational problems facing countries. The economic, organizational problems stem from the need to incentivize (motivate) and coordinate human activity to provoke productive activities. These activities occur in a constrained, riddled environment. The problem is to decipher the nature of the most optimal societal arrangements and understand the determinants of such arrangements.

The above discussion reveals that TS involves a *collective process* to reconfigure its constituent components and a *goal* to overcome market constraints to induce economic development, prosperity, and well-being. The arrangement of these constituent tourism components is complex. That is, achieving, and maintaining a reasonably significant competitive advantage requires synthesizing the necessary resources held by and available from various supply agents. Thus, as a destination organization vying for a market segment to benefit stakeholders shoulders the costs of research, development, production, branding, and marketing of that segment's foundation and inception. However, stakeholders (companies) not directly involved in the efforts to facilitate collaborative and trust structures, could well share in any advantages of the product for which they bore no costly responsibilities.

An added complexity includes the various supply stakeholders' possible moral hazard issues where they want to imply that their product is of a great quality when, in fact, it may be substandard. Such moral hazards are detrimental from a cost and trust perspective and do not bode well for the industry's overall welfare, and the destination's reputation. The destination's reputation depends on individual stakeholders' behavior, and stakeholders will behave responsibly only if their behavior is known or observable. Protecting the destination's reputation entails a private cost to the individual stakeholder, while all stakeholders comprising the destination share the benefit. If neither cost nor benefit are aligned, the individual stakeholder has incentive to freeride. Because the group's reputation is a collective good, an individual stakeholder has a weak incentive to behave well if the destination has a poor reputation.

Overall, tourist, and business and government institutions maintain some competition regarding the products' development from its initiation to its output. The problem materializes in need to balance these three entities' interests, particularly amidst the supply agent's moral hazard problems. That is, if supply agents cheat on the value and quality of the product, the tourists may likely deny the consumption of that product in that they cannot get what they feel they are entitled to, given the money they are paying for a memorable experience. In brief, their experience is substandard along with the product. Thus, if tourists are not consuming, the business is not profiting. If the business is not profiting, the government is not benefiting from tax revenues and increased employment opportunities for residents, leading to their well-being on and on. Sort of a "for want of the shoe, the horse was lost, for want of the horse, the rider was lost…" concept-eventually resulting in the loss of the kingdom.[17]

This possible destruction of the market would need an adequate remedy. Thus, Benjamin Franklin's proverb's wisdom reveals a need for positive steps to ensure that offerings are of the quality that visitors expect. Each member of the goods and services market must adhere to expectations to ensure repeat tourist visits. Tourist loyalty drives small businesses' incentives to continue to produce a quality service as there is no reason to believe the money the small business expends on that quality will return proportionate to tourist spending. Moreover, all product entities must coordinate strategically in a cooperative manner to ensure success from conception to development, branding, and presentation. Otherwise, to finish Franklin's proverb: "for want of the rider…the kingdom was lost."

The unique features of tourism as a service and experiential sector that trigger its consideration as an economic growth instrument stem from its demand characteristics. First, consumer preferences in high-income countries reveal income elastic tendencies towards recreational activities, including vacations. These tendencies imply that as income increases, consumers in high-income countries tend to spend a higher proportion of their income on vacationing while only slightly influenced by price variations. Once the many materials need and wants are satisfied, smaller amounts of marginal utility are obtained by additional consumption of the same products. The desire for something different is simply an outward manifestation of the marginal utility concept. The desire for travel has become a valuable source of additional marginal utility for the consumer with possible extenuating economic value to the tourism sectors itself.

As tourists seek memorable experiential opportunities, they examine destination culture and business venues that acquaint them with distinctive local character. They consume goods and services while at the venue

---

[17] Benjamin Franklin penned this proverb "For want of a nail" in 1758 in Poor Richard's Almanac.

and to take home-preserving their experience. Thus, their demand for local products is increased beyond that of the demand anticipated by the residents. Consequently, domestic supply becomes essential to meet the increased demand; prices can fluctuate following that demand without the burden of equating demand with terms of trade (ToT).

The ongoing supply and demand interactions in tourism propagate dynamic economic and social patterns. First, these interactions may provoke endogenous comparative advantage effects. Endogenous comparative advantage includes factors that are malleable to improvement or changes to suit the customers' tastes and preferences. This malleability stems from learning-by-doing (experience in the production process) and expanding demand for the good (demand elasticity).[18] Endogenous comparative advantage comprised of "those resource endowments that are more easily changed by endogenous factors, such as physical/human capital investment and technological innovation, in the correspondent country's economic system."[19] In other words, the endogenous growth of a sector has to do with forces within that segment (among others, leadership, creativity, innovation, and entrepreneurship) that command its outcome.[20]

Second, in terms of overall vacation costs, the tourism supply of goods could profile high elasticity. Some products' character cannot be substituted by other products, thus making them immutable. These products, such as a tropical rain forest, are not transposable, maintain a supply permanence in the market, and do not give way to other goods' production as substitutes. Some marketable goods, such as clean air, are considered non-rival, the supply of which is unaffected by public consumption. That is, available to all, this good can be used by many simultaneously and does not dissipate its utility or value according to individual demand. The goods and services that tourists require and want, encompassed in their reason for visiting a destination, can be met without negatively compromising overall tourism prices. That is, prices that a tourist pays for amenities and entertainment are not beyond those that the resident would pay.

Moreover, third, manufacturing goods do not seem to suffer from supply constraints compared to tourism. In general, natural resources are scarce, and over-crowdedness and a deterioration of natural assets pose a

---

[18] See, for example, Schianetz, Kavanagh, and Lockington (2007).

[19] See Hong (2008), p. 54.

[20] Lanza argues that the "...knowledge produced in the economy specializing in the manufacturing sector cannot be 'transferred' to tourism, given the apparent difference existing between the two productive sectors and the nature of tourism, which is concerned with providing services to people...."

challenge to policymakers. Therefore, Lanza suggests two conditions for TS to trigger economic growth: income elasticity and supply constraints. The value that a tourist attaches to a product anchor on its usefulness. Because utility may vary between individuals and at different times, value assumes a subjective character-driven by tourist preferences. Tourist preferences typically bind supply constraints in the case of tourism. These may reveal income elasticities, for example, beach tourism; however, the magnitude of this demand is limited because beaches and corals are exogenously determined, and there is no possibility to increase their supply.[21] The result is that the destination value is determined by the concepts of scarcity and preferences of tourists.

2.5   THE TOURISM SPECIALIZATION (TS) HYPOTHESIS

In tourism, each person is a consumer/producer due to the simultaneous consumption and production process in the same location. For tourism, a primary determinant in the production of goods and services is the tourist himself. In other words, it is the tourist who dictates how the production of goods and services moves in real-time. Suppose the structures that bear an impact on the tourist industry (government, social mores, etc.), are not in complement or compatible with tourist preferences. In that case, demand dissipates, and opportunities to use the tourism sector as a means for economic growth could be compromised or lost forever.

The simultaneity of consumption and production in the tourism process implies that an increase in the division of labor is interpreted as an increase in the proportion of an individual's consumption that is purchased. The purchase of a tourism product involves a transaction cost. Inferring from Smith's logic about the division of labor that all individuals are identical (no natural differences), we may conclude that for an exchange to take place, individuals should display a positive relationship between labor productivity and transaction costs. Therefore, the problem does not seem to be the efficient allocation of resources; instead, the problem is the choice about the level of specialization and the consumption variety that goes with it. This choice is deliberate and hinges upon societal learning.

---

[21] It is crucial to invest in the preservation of the natural resources and to increase product differentiation, in particular, due to the monopolistically competitive market structure of international tourism. For an interesting discussion of destinations' interactions promoting product differentiation, see, among others, Candela and Cellini (2006).

This societal learning depends on how residents feel and perceive the benefits prompted by TS.[22] Whether objective or subjective, these benefits depend on the resources generated by tourism activity available to residents. Also, residents' feelings are affected by the opportunity to use these resources to achieve valuable life conditions, such as life expectancy, literacy rate, incomes, and life satisfaction and happiness. Feelings and perceptions may be subject to bias due to hedonic adaptation. For example, this bias, which colors the measurement of subjective well-being, is subsumed in an error term for any judgment of a person's life situation shaped by tourism development.[23] The process is dynamic because the supply and demand conditions are ever changing, which induces and infuses constant societal learning. Tourism production reveals a simultaneous interface between supply factors and demand, and tourism economic prowess relies on demand elasticity.

In tourism, these two conditions are not automatic. As mentioned previously, the problem manifests itself in incentive and coordination problems present at a destination.[24] Several commentators demonstrated that income effects vary over time and shape tourism consumption. These income effects may be due to two unique tourism production and consumption features.[25] Tourism is a perishable activity (an unoccupied hotel room not sold is lost forever). Also, the tourism activity links the production and consumption both temporarily and spatially, restricting rationalization opportunities. The restriction of these rationalization opportunities means that the costs of engaging in the tourism business may increase over time because productivity in tourism jobs lags productivity in manufacturing.

Tourism peril lies in the reality that tourism products become more expensive than other products over time, and therefore may not be an engine of growth. The production costs of tourism may grow over time, thereby inducing higher prices with less demand. This is the so-called Baumol's cost disease. This disease entails that tourism productivity is lower than in other economic sectors, while wages in tourism increase as fast as or faster than wages in other sectors.[26] Productivity growth must

---

[22] For a thorough discussion on this learning process induced by tourism specialization, see Nowak et al. (2010); and Marsiglio (2018).

[23] For an interesting discussion on this subject, see, for example, Croes (2012).

[24] For a discussion on these coordination problems, see, for example, Candela and Figini (2010).

[25] See, for example, Smeral (2012, 2016); and Croes and Ridderstaat (2017).

[26] This cost disease can affect the destination competitiveness, because its slacking productivity growth compared to other sectors can only be compensated through higher prices. However, there are several ways to overcome the cost disease peril (see Smeral, 2003).

be higher than price growth to maintain the profitability of tourism. Of course, this situation could cause welfare reduction. Endogenous growth theory regards growth only through the eyes of high productivity sectors. However, empirical evidence seems to contradict this pessimistic look as TS has generated fast and high growth rates, particularly in small island destinations.[27] However, the critical question is, for how long?

The advantage seems not to attain a cost advantage in tourism production, but in extracting higher prices over time by increasing demand via increased product appeal. This appeal involves tourists' working tastes and preferences so that tourists' goods are increasingly valued in international markets, thereby offsetting the increase in cost production to raise or maintain product prices under conditions of increased competition. Small islands could create more opportunities where locals and tourists can interact to provide an authentic flavor of the local culture. The interactions are timely and following the synapse between the actors and the physical and social arena in which that synapse occurs. While heritage, cultures, and geographic planes may be unchanging, the interactions between the locals and the tourists are as inimitable and distinctive as the actors are diverse.

The TS hypothesis follows Popperian premises. The properties of TS are to be discovered; they exist in the world of facts (reality). The construct consists of specific properties (i.e., process, complexity, dynamism, and outcome). However, the construct is a human creation. As we showed in the previous sections, the construct of specialization has a long provenance. The construct is also consequential because the world would be different without it. Specialization features prominently in economic development debates, discussions about urban growth, and economic growth in small island countries and developing countries. Consider tourism, for example. Without TS, millions of people would lose their jobs; small island economies would become saddled with economic depression and poverty. The TS hypothesis anchors these Popperian premises in the TS hypothesis. The hypothesis follows the Lucas endogenous growth model's logic to establish the relationship between TS and economic growth. In the Lucas model, the ratio of human capital to physical capital determines the

---

[27]Numerous studies examined theoretically and empirically the connection between tourism specialization and economic growth, anchored in the TLGH framework. Several theoretical works demonstrated that tourism demand is characterized by a low elasticity of substitution, which implies that tourism may effectively promote growth only when the terms of trade move in favor of tourism activities (Lanza, Temple, and Urga, 2003; Brau, Lanza, and Pigliaru, 2007; Schubert, Brida, and Risso 2011).

dynamic economic intensity. This ratio has increasing return effects when in an open economy, which implies that trade is an integral component of the economy.

Once the choice is to export, the model opens up opportunities for specialization, which leads to higher productivity and increased well-being. This choice cannot be determined a priori and its effects depend on what and how residents learn from this specialization experience. The production of externality's consequential effects can be either positive or negative, revealing short-term or long-term trends. The channel through which specialization promotes economic growth in tourism stems mainly through demand-side factors (income elasticities) and supply-side factors (learning-by-doing). In a TS scenario, the relationship between physical and human capital becomes more important in economic production, signaling a favorable scenario when the shift is towards more human capital-intensive production; or economic production may swing to an adverse scenario when production becomes less human capital-intensive.

Figure 2.2 illustrates the TS hypothesis framework in small islands.

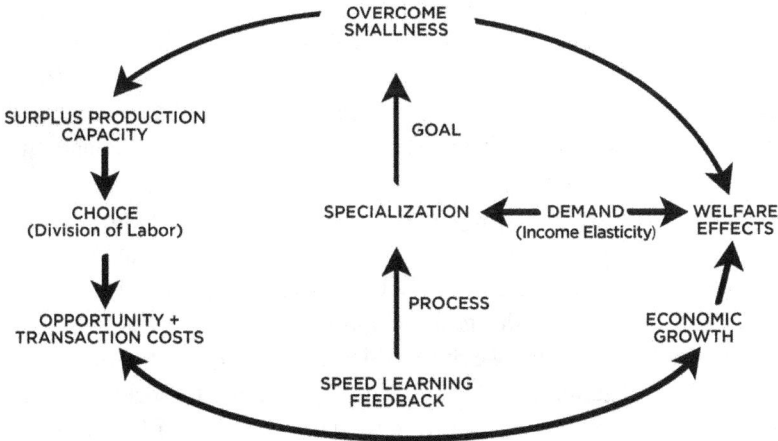

FIGURE 2.2   The tourism specialization hypothesis and smallness.

2.6   DOES SPECIALIZATION MATTER FOR SMALL ISLANDS?

Small islands seem particularly prone to engage in TS because small islands entertain a smaller opportunity cost of specialization. TS can be a determinant of economic growth. However, the literature explaining the relationship between tourism and growth has been mixed so far. Ascertaining whether tourism bears economic growth a priori is neither feasible nor attainable. The literature indicates that tourism costs are prohibitive, thus rendering an a priori assertion prohibitive as well. Indeed, according to Smeral's study, there are indications that the cost of the tourism service has surpassed the cost of other destination goods. In his study, Smeral cites tourist consumption and destination productions as the two main reasons for this. First, it must be understood that tourism is a perishable entity. If the auspices provided for tourist experiences, such as accommodations, are not sold, they are lost. Moreover, second, the connectedness between tourist consumption and goods and services production deprives tourist businesses of the due diligence required to rationalize and reconcile production to control costs.[28]

If the cost of goods and services continues to surpass other goods, which could happen if costs continue to rise over time, as Smeral posits, tourism will cease to contribute to economic growth-in part because tourism jobs' productivity falls behind that of the manufacturing sector. This lag does not lead to or contribute to economic growth in the long run if one supposes the endogenous growth theory, at which point growth is forwarded by sectors emanating high productivity degrees. Yet, from an empirical standpoint, small island destinations where TS is fervent indicates these high productivity degrees. Moreover, small island destinations are subjected to specialization costs that tally less.

Where small islands employ their natural assets, such as Sun, Sand, and Sea, or cultural heritage sites, and where the tourism vs. manufacturing productivity is not too proximal, ToT can neutralize the consequence of the productivity gap. In the tourism industry, where goods maintain an inelastic supply, the response amid a demand that quickly fluctuates can impact how production processes are strategized and how creative human agency and developmental, future potential move forward. The resulting reaction implies that the scarcity of tourism products could force higher rents to consumers.

Forced higher rents could also be a consequence of how tourism goods and services are consumed. Research indicates that tourism is characterized

---

[28] For example, tourism has positioned itself as an important consumer of public resources, potentially generating imbalances in islands' budgets affecting critical areas such as health, education, or social care. I am indebted to Eduardo Parra-Lopez for this point.

as nonessential or of a luxury nature insofar as its demand indicates an income elasticity beyond 1. This continuing characterization reveals less budgetary competition between tourism and non-tourism goods and services for individuals needing to deploy their available financial assets to essential vs. nonessential products and services. This being the case, pricing for forcing higher rents to consumers to counter the characterization could be more successful when the tourism base is more unique and diversified. Any deficit, then, might be alleviated via specialization costs that favor the destination.

The greater the demand for goods and services that tourists proffer, the more the tourist sector becomes consigned to economic growth. This occurs in three ways. First, tourists allocate their money for accommodations and amenities that include hotels, entertainment, cultural events, visits to heritage sites, etc. Profits collected from such venues are portioned directly to the businesses, employee, and resident households, and the government in taxes, licenses, etc. Second, tourist demand poses an ancillary impact on GDP via the spillover that occurs when new ideas and new businesses establish themselves at the destination.

As these innovative management and development structures and skilled labor and professionals enter the industry on-site, new products and services also enter the tourism market. Subsequently, having learned and been inspired by new establishments, local businesses might better meet the needs of tourists by revisiting their standards of product quality and service. Moreover, third, ToT may compensate for the productivity gap mainly because of the comparative supply constraints. These constraints provoke, comparatively, an inelastic reaction of supply to rapidly changing demand, which enables a destination to extract higher rents because of scarcity.

As economic growth, potential job availability, and entrepreneurial benefits continue to perpetuate the evolution of tourism as a viable path to prosperity, countries of the Mediterranean and small islands such as the Caribbean developed critical strategies to secure their factors of production, thus working towards seizing and maintaining their market share. However, the standard features in tourism production are greatly overshadowed by the many economic differences among the Caribbean, the Mediterranean, and the Pacific regarding their natural endowments, economic performance, wealth, and well-being. Small islands seem particularly prone to engage in TS because they entertain a smaller opportunity cost of specialization. It is convenient for smaller sized islands to specialize in tourism. Additionally, the combination of two elasticity phenomena (high-income elasticity and price elasticity ambiguity) contributes to relatively stable export earnings of tourism products compared to commodity groups benefiting the ToT of destinations specializing in tourism.

However, while tourism demand studies have played a prominent role in identifying the determinants of tourism flows, these studies have not provided sufficient empirical evidence regarding TS sustainability. Tourism development may be a temporary occurrence. Its development may not be sustainable, or tourism's impact may change over time based on tourism's developmental stages, or demand elasticities may fluctuate over time. The reason for the temporal issues related to tourism may be that growth is induced by the intensive and increasing use of one production factor in small islands: natural resources.[29] Once this factor's employment reaches its maximum use, labor productivity becomes crucial in determining growth, and, as a result, tourism countries may grow more slowly than others. From this perspective, tourism may not be a viable development path in the long-term.

Based on the previous discussion, this book conceives TS as a process consisting of four main premises: (1) production factors are inputs in the production process that generates utility (outputs) to the tourists; this production process rests on instrumental activities governed by rational behavior (costs and benefits analysis) and transaction costs generating economic value. (2) The tourism product depends on scarce resources implying that these depletable resources should economize over time. (3) Diminishing or increasing returns due to TS relies on the quality of the product and the simultaneous benefits accrued to residents and tourists. Moreover, (4) returns can change over time. The outcome of these premises' interdependence depends on three conditions, i.e., the demand elasticity (income elastic), the price elasticity, and how quickly a society can learn. This complex causal relation between tourists' preferences and supply quality underlies the TS notion of cumulative causation, which stresses sequential change and the progressive nature of these changes. As long as the demand is income elastic, the expectation is a positive ToT.

As small islands seem to suffer from scale hardships, they have to compensate for their small domestic market by engaging in trade with

---

[29] If differentiation in the tourism market relies on the quality of natural resources. Cerina (2007) and Marsiglio (2015) show that sustainable tourism through investment in environmental maintenance is critical for long-term growth.

the international market. We already noticed that trade is a growth avenue for small countries, as suggested by Kuznets. Several commentators also followed this line of reasoning-encouraging small countries and, by logic, small islands, to engage in international trade. Engaging in trade, however, makes them also more vulnerable to sudden shifts in international demand. Some small countries tend to be significant recipients of aid, preferential treatment, and remittances to mitigate their vulnerability. In the Pacific area, islands practiced this economic model to sustain themselves economically and socially. These islands got the name of MIRAB countries, which stand for migration (MI), remittance (R), foreign aid (A), and public bureaucracy (B).[30]

Alternatively, small islands also revealed different trade patterns. These different trade patterns result from their engagement with international finance and tourism. Both international finance and tourism integrated these islands into the global economy. Openness to trade has the potential to solve some of the persistent problems of scale economies. This potential rests on channels that connect economic growth with promptly improved resource allocation, increased competition, the quick achievement of knowledge and innovation, and reduced rent-seeking and corruption levels. These small islands uncovered substantial growth opportunities and have been depicting high-income growth levels and rates. Their growth trajectories suggest that specialization matters for growth in the case of small islands. The following chapters will systematically assess the empirical reality of the TS hypothesis.

KEYWORDS

- **dynamic comparative advantage**
- **economic development**
- **economy**
- **immobility**
- **specialization goal**
- **sustainable economy**

---

[30] See, for example, Bertram (1986); and Connell and Conway (2000).

CHAPTER 3

THE TOURISM DEVELOPMENT MODEL

The most effective means for achieving economic performance in a country has been the subject of considerable study and debate. Theories focus on such issues as a comparative and competitive advantage, openness of an economy, and economic scale hurdles and scope affecting economic performance. Studies on whether small size affects the economic performance have been numerous. For example, successful tourism development has been realized by Mauritius, an island on the verge of economic collapse. Previous to its impending collapse, Mauritius had been an agricultural society but suffered from the limited availability of national resources and its domestic market's inadequacy.[1]

The island currently continues to develop its tourism industry. Over 40 years, Mauritius has achieved rewarding economic independence, and enjoys a very high human development index (HDI). The island strategically focused on tourist market segments rather than considering and engaging strategies concerned with tourist volume as a corridor to economic feasibility. By developing a means to reduce its leakages, Mauritius has placed itself on a path to prosperity.[2] Indeed, the island magnetizes the type of tourist willing to consume its offerings. Moreover, in its course to develop and sustain its tourism industry, it has become one of the most preferred destinations in Africa.

For years, the literature seemed to consider smallness as a liability. Many scholars attributed the disadvantages that small size imposes on the growth process to the difficulty of achieving sufficient economies of scale in a wide variety of activities. In 1965, Demas argued that the comparative

---

[1] See, for example, Subramanian and Roy (2003). Meade's reading of the economic possibilities turned out not to be correct as Mauritius, over time, resulted in a success story in Africa. See on this matter Subramanian and Roy (2003).

[2] See, for example, Durbarry (2002); and Subramanian and Roy (2003).

disadvantage in labor-intensive manufacturing, for example, implies that alternative sources of growth that are less dependent on industrialization are needed. The inherent restrictions related to market size, resources, labor, and capital, therefore, force small countries to look to the global economy to achieve economies of scale in exports. According to Winters and Martin's study in 2004, smallness does not bode well for business organizations seeking to establish themselves in that type of locale. Business organizations would risk shocks if their return on investment were negative. Further, the products they seek to sell in an open global market would fail to pursue any competitive advantage.[3]

Several concerns underlie the pessimism about the opportunities and possibilities of small islands to engage in the development activities that could contribute to and carry the destination forward. Included here are the challenges of compensating for scale disadvantages. The ability to sidestep scale requires small island destinations to connect with potential growth and development opportunities that lie outside the island borders. More often than not, small island destinations have embraced trade as the island's potential for economic growth and development. Thus, the potency and impact of trade on a small economy, especially one that is confined, is substantially more extensive than that of other market economies. It has been empirically estimated, for example, that just one standard deviation in the degree for openness increases the growth rate of a smaller country by more than three times that of a larger country.[4] Thus, the impact of trade on smaller country economies has led to relatively high growth rates. However, as small island countries depend upon trade to bolster economic gains, dependency upon trade markets impacted by oscillating supply and demand can lead to trade shocks that increase volatility in a country's GDP more so than in larger countries.[5]

Because small island destinations are limited in commodities and resources and available export markets, they experience increased trade volatility. As islands tend to be agriculturally based, prices for agricultural products fluctuate by supply and demand. This fluctuation can negatively

---

[3] For example, Sir Alister McIntyre noted that the average cost of producing sugar in the Caribbean was US$ 535 per ton compared to US$ 266 in the Pacific and US$ 340 in Africa; for bananas, the average in Jamaica was US$ 391, whereas it was US$ 291 in Colombia, US$ 179 in Costa Rica and US$ 161 in Ecuador (McIntyre as cited in Griffith, 2002).
[4] See, for example, Helpman (2004) assessed the effects of trade on economic growth comparing Mali and Seychelles.
[5] See, for example, McGillivray et al. (2010).

impact the islands' economic performance. This instability has been observed in the Pacific islands' national income.[6] In general, high resource-based economies tend to grow slower over time because they are less high-skill labor intensive with little spill-over effects to other economic sectors. The accumulation of adequate production factors (physical capital and human capital) has proven to be challenging for these countries. These challenges, combined with risks of resource depletion, the decline in demand, and technological change and requirements of less human capital and skills, may impact the economy. Commentators also claim that globalization forces have rendered small islands more vulnerable due to their inability to deflect the consequences of volatility and their dependence on a small pool of international markets. Volatility propagates unemployment, lack of consistent investment, risks, and instability in the socio-economic fabric.

3.1   DUBIOUS MERIT OF THE TOURISM DEVELOPMENT PATHWAY

Traditionally, manufacturing has been considered a rational and reliable activity to perpetuate economic growth. Manufacturing has the capacity of economies of scale and its efficiency can reduce production and consumer costs, indulge in increasing demand, and exploit the export of goods. This perspective does not bode well for the tourism industry. Instead of an institution with the potential to realize economies of scale, it is characterized as grossly restricted in its ability to fruitfully participate in trade on a global scale given that demand for its goods and offerings are confined to and consumed on location. Accordingly, viewing global trade prospects in this manner implies that growth can only occur as long as domestic consumers indicate a demand-a short-lived prospect indeed.

Overall, then, the tourism literature has not been kind to small island endeavors to secure their development via the tourism industry. In his 1989 study, Wilkinson derided small islands practicing tourism development based on the SSS model as "folly." Tourism development as an unsuitable development strategy for small islands was a popular proposition among social scientists and international organizations. Moreover, there seems so much resistance in the literature against the idea that small islands could

---

[6] See, for example, Thirlwall (1991); and Santos-Paulino (2010).

thrive, prosper, and provide a decent quality of life to their residents. Many skeptics characterized tourism development that could lead to prosperity and enhanced living standards in small countries as wishful thinking. However, why does this negative perception of tourism development exist and persist? Patullo's *Last resorts: the cost of tourism in the Caribbean*, for example, could provide the answer to this question.

Patullo's study suggests that tourism incurs considerable severe social, cultural, and environmental costs. For example, this study alleges that tourism is welfare-decreasing because it displaces agriculture to tourism labor without benefits. The study characterizes this displacement in pejorative terms-farmers moving as "banana farmers to banana daiquiri." The study also dwells on other claims such as tourism spawning social costs revealed in white exclusivity, an allusion to slavery, prostitution, drugs, and crime, and environmental costs uncovered in the destruction of shoreline vegetation, wetlands, and salt ponds, coastal erosion, sand mining, beach erosion, reef damage, etc.

Woefully, but often deservedly, tourists are blamed for being carefree and careless about the local culture, and uninterested in interacting with the local population or in learning about its traditions and norms, showing interests only in the pleasures of Sun, Sand, and Sea. Destinations practicing mass tourism are perceived as inferior to other places or to what the place used to be like. Mass tourism seems to have troubled small islands. One lurking example of this is the case of the Balearic Islands. The Balearic Islands were viewed as the epitome of mass tourism revealing a path that SSS destinations would follow, illustrating high environmental costs and decline. These islands were equated with tourism running amok, a powerful example of out-of-control development that values short-term profit over sustainability. The Balearic Islands got a nickname for their run-amok tourism development: *balearización*.[7]

Another example is the case of Malta. In Malta, the extensive growth of the tourism industry has served to provoke the residents.[8] Malta's tourism industry evolved rapidly in the 1980s, built mainly as a SSS product and characterized by a mass tourism infrastructure. The rapid tourism development on Malta altered the overall general positive disposition to tourism and evinced negative sentiments from some groups demanding more attention to the environment and better destination planning and management.

---

[7] Mass tourism and balearización seem the precursor of the phenomenon overtourism.
[8] See, for example, Bramwell (2003).

Boissevain and Theuma, in their study published in 1998, refer to these attitude changes as follows: "Until the mid-1980s, most Maltese unreservedly
welcomed tourists. They accepted that maximizing tourist arrivals and the
resulting overcrowding, discomfort, rampant building, and environmental
destruction was necessary for economic development...In the 1990s, as
tourists topped 1 million annually, the Maltese began to feel oppressed by
the effects of this laissez-faire pressure on the social and physical environment... Recent public protest about threats to Malta's environment have
all concerned new projects aimed at attracting upmarket tourists." (pp.
97–99). The vast tourist numbers seeking SSS thus presented a toll too
great on the sustainability of the industry.

Critics gave little regard to leisure travel, even attacking it as nearly
destructive to societies that practice tourism. In his study published in
1982, Britton emphasized the tourism degenerative nature due to uncon-
trolled growth and overexploitation of natural and cultural resources.
Further culpability to the degenerative nature was mostly ascribed to
neo-colonialism, foreign dominance of multinationals which promote
economic leakages and repatriated their profits to their headquarters
located outside the small island, and the extant dependency mindset of
local island residents acting as victims and lacking the creative agency to
change their dependent course and to make the necessary adaptations and
adjustments toward independent control. Goldstone in her book published
in 2001 described tourism as an extremely vulnerable and fickle activity,
spawning social division, fostering intimate relations with organized
crime, and serving the interest of repressive regimes and multinationals.[9]

The studies of Bertram and McElroy, published in 2004 and 2006, also
echoed these negative sentiments regarding tourism's role in the developmental quest of small islands. Skeptics used tourism's moralization as an
argument against harnessing tourism as a transformative power for small
islands.[10] The whole notion of the SSS development model, perceived
as mass tourism, was under assault-portrayed as essentially the destructive nature of economic development.[11] Critics also claimed that tourism
development makes small islands less competitive because mass tourism
is unsustainable. According to Butcher, "... [tourism] is increasingly
discussed less as an economic phenomenon linked to the creation of jobs

---

[9] See Goldstone (2001).
[10] See, for example, Patullo (1996).
[11] See, for example, Butcher (2003).

and investment, or indeed simply as enjoyment, adventure, and innocent fun. Rather, tourism has increasingly become discussed as cultural and environmental phenomenon … as fraught and destructive" (p. 6), considering tourism and nature as mutually contradictory.[12] According to critics, tourism should be founded on nature conservation rather than the transformation of natural resources.[13]

Multilateral organizations, such as the World Bank, revealed resistance towards tourism development for small islands. Such a situation has arisen in the Caribbean. As development advanced, and the islands focused on production, negative characteristics emerged, and the potential competitiveness that the Caribbean sought to control diminished. As a result, the Caribbean's endeavor to sustain itself as an apex tourist destination has depleted.[14] In another study, Hawkins, and Mann chronicled in 2007 how the World Bank disengaged from tourism as a developmental tool and focused its efforts on micro-projects from 2000 onwards. The lack of interest in tourism by the World Bank seems a result of tourism being perceived as "…unstable and volatile, with destinations at the mercy of trends and fashions for their popularity, dependent upon fluctuating political and economic conditions worldwide, and impacted by natural/human-made disasters and political instability." (p. 353). The World Bank was outright critical to the merits of tourism development as a self-sustaining economic activity.

3.2 THE LEAKAGE TOURISM PROBLEM

We already noted critical observations from several commentators not enthralled with tourism prospects and who portrayed tourism as an activity incurring high opportunity costs. Tourism specialization (TS) may involve high opportunity costs of resource allocation and has therefore been dismissed as a viable economic development strategy for small countries. A tourism productivity gap spurs the main reason for this concern with opportunity costs. This productivity gap makes additional inputs of capital and labor costly, and these additional inputs would not necessarily lead to

---

[12] See Butcher (2003).
[13] This perspective seems to be grounded in the United Nations Rio Earth Summit in 1992. This perspective seems to have removed people as the central part of nature, and hence, has limited the transformative power to harness and organize nature to provide for people's needs and wants.
[14] See, for example, World Bank (2005), (p. 100).

the desired growth level. There is even concern regarding tourism development imbued with diminishing returns.[15] I will argue against this thought in Chapter 7, claiming that terms of trade (ToT) can offset the productivity gap, thereby inducing lower opportunity costs of specialization.

A common criticism lodged against tourism development in small islands is the leakage phenomenon. Leakage is the money that a tourist spends to visit a destination, but which never enters that destination's economy. Acquired monies derived from tourist spending flow externally to cover expenditures that tourism development requires to grow and progress. Then, these expenditures constitute external leakages of necessary monetary assets as they route to off-island sources. Unfortunately, such sources are crucial to the successful foundation of the island's budding tourism industry. They can be of considerable impact on the small island's available capital to sustain itself in a manner that allows tourist expenditures to course more wholly into the destination's economy.

Sources can include foreign financial input earmarked for the destination's supportive infrastructure, including construction costs and facility upgrades. Crucially, ongoing amortization of existing foreign debts-the profits of which do not flow back to the island destination. Also, imports of foreign goods and services, transportation (including bookings, cruises, and airlines) costs and arrangements, advertising programs to foreign markets are fluid costs that contribute to external leakage. Moreover, the leakage phenomenon is burdensome to the small island destinations' overall practicability, potentiality, and probability for ongoing, independent success.

Gossling referenced this problem in 2003 and alluded to the high proportion of money staying outside of the destination's economy, especially in a small island context. Gossling's claim regarding the high leakage follows Wilkinson, and Wilkinson followed the study conducted by Seward and Spinrad in 1982. In their study published in 1998, Mowforth, and Munt also mentioned leakage as a problem for small islands and alleged this leakage to be spurred by the unequal power relations between these small islands and transnational corporations. For example, they cite Patullo's study that the leakage problem in the Caribbean is, on average, 70%. Both Mowforth and Munt, and Patullo's studies are based on the estimation from the Seward and Spinrad 1982 study.[16]

---

[15] See, for example, Amadou and Clerides (2010).
[16] Seward and Spinrad (1982).

Because the latter study seems to be the basis of later references to tourism, causing high leakage in the Caribbean, I had a closer look at the Seward and Spinrad study. I focused specifically on the Aruba estimation due to data availability and my familiarity with the destination. The estimation of tourism leakage in Aruba by Seward and Spinrad suffers from several drawbacks. First, the authors are vague about the materials used in their calculations. For example, the authors mention on p. 108 that they used hotel financial statements and observation to estimate foreign exchange earnings for 1980. However, they are not specific about how many financial statements they reviewed (all? one?) or what their observation entailed. Also, on p. 110, the authors indicate that they used additional research-based assumptions to distinguish between local and foreign expenditures and estimate the second-round leakages, but do not enlighten the reader about their conventions. These nebulous descriptions do not contribute to the credibility of the authors' estimations.

Second, the expense component's estimation as a percentage of operating income is peculiar, given that the overall (Total = sum of all expenses) is set at 100%. If companies make profits, the obvious conclusion should be that the profits are larger than the costs (expenses), and linking expenses to a higher denominator (i.e., profits) would, in the end, not result in 100% when considering the total of all expenses. Analysis of the last footnote of Tables 7, 8, and 10 of the study of Seward and Spinrad indicate that the fixed expenses component includes profits, which is an erroneous assumption and invalidates the calculated percentages and the derived calculations in these tables. Continuing with the fixed expense component, this segment also includes depreciation, which is an expense but involves no cash outflow. This is important because this implies that depreciation would have no leakage effect. Depending on the applied methodology, length of depreciation, and type of depreciated asset (e.g., buildings and cars), the depreciation component could be substantial per year.

Third, the authors implicitly assume that locals' imports are solely attributed to foreign exchange received from tourism. In those years, the island's economy was influenced by an oil refinery (The Lago Oil and Transport Company, Ltd.) that contributed to between 30% and 50% of all foreign exchange receipts. Considering all import-related consumptive spending by households as part of the second-round leakage calculation causes an overstatement of the latter. And fourth, the authors consider purchases of foreign-made capital goods for expansion or improvement of facilities as part of the leakage. However, spending is often funded by capital inflows instead of tourism receipts. These capital inflows are not recorded in a country's balance of payments as tourism receipts but as foreign investments. These outflows should not be part of the second-order leakage.

As suggested earlier, tourism studies have in general uncritically embraced the Seward and Spinrad study's tenet of high leakages. My analysis shows it is likely that the study by Seward and Spinrad may have overstated the estimated tourism-related foreign exchange leakage. Unlike this study, Lejarraga, and Walkenhorst,[17] based on the World Travel and Tourism Council satellite accounts, suggest in 2007 that leakages in the Caribbean are between 30% and 40%. In my Small Island Paradox published in 2011, I estimated the leakage for Caribbean islands in the range of the estimates provided by Lejarraga and Walkenhorst's study. For example, the leakage estimate for St. Vincent and the Grenadines was 0.35; Grenada was 0.34; St. Lucia was 0.29; Barbados 0.38, and Dominica was 0.30. Only St. Kitts and Nevis scored 0.51.

3.3   AN OPTIMISTIC PERSPECTIVE ABOUT TOURISM DEVELOPMENT

However, is it true that small islands are practicing mass tourism pernicious to their prosperity and well-being as these commentators want you to believe? Small islands have been depicted as vulnerable and non-viable with limited development prospects. I do not share this deterministic and pessimistic picture of tourism in the Caribbean. These claims arise from stereotyping tourism development in the region and are biased towards its prospects. Stereotyping small islands leans their potential toward globalization triggering vulnerability, rather than toward the positive prospects the islands could or would otherwise embrace.

Small islands suffer from the trappings of labels. Usually, small islands are viewed out of context or out of real time evidence. Such is the characterization of small islands as highly globalized. A study from Dreher published in 2006 shows that small islands such as the Bahamas and Barbados are low in globalization. Dreher constructed a weighted

---

[17] See Lejarraga and Walkenhorst (2007).

globalization index on flows of goods, capital, services, information, and restrictions on trade and capital flow. His study characterized globalization as a multidimensional construct consisting of economic, political, and social integration using 23 variables to measure these dimensions from 1970 to 2000. The globalization index ranges from a zero to a 10 scale with higher values denoting more globalization.

The results of this constructed index revealed that small islands, contrary to expectations, were not highly integrated with the global world. This result persisted in the most recent global ranking. For example, the Bahamas and Barbados ranked at the bottom with scores of respectively 101 and 104 in a list of 114 countries. Table 3.1 illustrates the globalization ranking.

TABLE 3.1   2010 Globalization Ranking of Selected Islands

Country	Overall	Economic	Social	Political
The Bahamas	51.61	35.09	77	30.9
Barbados	59.72	51.22	80.23	45.5
Cyprus	81.35	79.48	87.27	77.29
Dominica	46.07	55.62	70.27	31.71
Fiji		51.61	66	51.34
Grenada	50.28	49.2	69.9	43.4
Jamaica	66.75	60.71	69.08	67.47
Malta	77.26	87.41	84.16	60.2
Mauritius	68.97	81.58	71.99	53.33
St. Kitts	46.95	53.41	80.8	20.13
St. Vincent	45.78	50.86	68.24	30.82
St. Lucia	48.93	57.85	72.58	36.28
Trinidad and Tobago	62.05	62.98	66.2	55.76

Akin to my viewpoint, Scheyvens and Momson in 2008 also contested the pessimistic look regarding small island development prospects. Their study suggest that small islands are appealing to international tourists as exotic places displaying several strengths that can positively undergird tourism development. This appeal endorses the growing demand to visit small islands. I have seen and experienced small island destinations engaging with volumes of arrivals while simultaneously understanding and practicing segmentation. Small island destinations have shown that they understand that motivations within niche segments can vary. There has existed a need to understand these motivations about having fun, shopping, visiting friends and relatives, visiting sites, engaging in cultural events and activities, or simply relaxing. Indeed, international tourism to small islands has grown faster than the world's average from 1990 to 2002. The study concludes: *"In light of the evidence presented above, it appears that the vulnerability and dependence of small island states have been overstated in much of the tourism and general development literature"* (p. 505).

Butcher's "The moralization of tourism. Sun, sand… and saving the world" pushed back on the role of tourism as simply package tourism by dubbing this criticism "moralization of tourism." Butcher asserts that the comparisons between mass tourism and alternative tourism types, including ecotourism, are inadequate, misplaced, and misleading, value-laden critique. Butcher asserts that this moralization is anchored on two notions: first, mass tourism is bad for a destination's environment and culture. Second, a new tourist should replace the old tourist because the former is benign to the environment and benevolent to the local culture. He posits that mass tourism could occur naturally in moralistic tourism, given that the destination does not change its resources to accommodate the traveler.

Thus, though the rationale for visiting a destination may contrast visitor to visitor, there are still static elements of the destination that exist and are available to all despite contrasting visit purposes (e.g., they would access the same transportation, hotels). By way of example, consider the cultural tourist. At the same time, his visit may be driven by the opportunity to engage with residents, their music, foods, and customs. This is not to say that the mass tourist would shun the opportunity to participate in these same offerings in spite of his primary purpose, which may be to photograph the historical sites.

A comparison of moralistic tourism to mass tourism could also be made similarly-occurring naturally. For example, the cultural visitor (moralistic) may well choose to participate in the same activity as the mass tourist. All tourists can be mass tourists since they are part of this mass leisure phenomenon and vice versa. The destinations and their industry should be seen rather dynamically, adopting features from new tourism, for example, through different activities. Mass tourism in the Mediterranean has expanded 'beyond the beach' by 'diversification of the tourist experience.'

The criticism of mass tourism also implies that SSS destinations have experienced reduced competitiveness. Does this claim bear out, based on empirical evidence? According to Aguilo, Alegre, and Sard's article in Tourism Management, the empirical evidence in the Balearic Islands does not support this claim. Contrary to what the critics of the SSS model have been preaching, this study reveals that tourists visiting the Balearic Islands invoke several motivations, including beaches, fun, quality hotels and surroundings, relaxation, and proximity to main source markets, portraying these islands as more than just Sun, Sand, and Sea destinations. They also chronicle how the islands have restructured the markets to remain competitive by specializing, segmenting, renovating, and beautifying urban areas and accommodation facilities. The study concludes that, despite doubts, SSS destinations have persisted.[18]

Torres in 2002 examined mass tourism in Cancun and the surrounding state of Quintana Roo. Her empirical study drew data via survey from 615 visitors and 60 hotels to determine whether the destination conformed to the Fordism and the Post Fordism spectrum or both. Considered initially to be a destination in character with the Fordist concept, Torres indicated the region to be both Fordist and Post Fordist. As populations and global interactions diversify, so too does the movement and combinations of Fordism, Post Fordism, and Neo Fordism commence. Moreover, there appears a more varied tourist image that stems from their motivation to visit a given destination, which then contributes to the destination's overall image.

With Manuel Vanegas, I also embarked on the path to assess tourism's developmental role in Aruba. Aruba's tourism development experience by 2002 seemed to contradict the pessimistic notion regarding tourism development in small island destinations. Aruba has considerably expanded its tourism infrastructure, witnessed strong growth in international arrivals and receipts, saw substantial economic growth, and tripled its real income per capita. It has shifted from the oil industry to a service industry, and in the process, reducing the nearly 25% unemployment to full employment within 10 years. While embracing the SSS model, its quick accomplishments suggest a strategy oriented towards a SSS quality plus model.[19]

My 2006 study, *A paradigm shift to a new strategy for small island economies*, also discussed why tourism development might assist small islands' developmental goals. In this study, I proposed and examined a

---

[18] See also Buswell (2011).
[19] See Croes and Vanegas (2003).

strategy anchored on a SSS quality plus model. The model shifted away from the conventional supply-side model to a demand-enhancing model based on micro segmentation and revenue optimization that focused on a better understanding of tourists' tastes and preferences (preference maximization). This model is premised on three interrelated claims. First, the value enhancement of a destination is more sustainable than more tourism arrivals. Second, successful TS hinges on the low elasticity of substitution of tourism goods and other non-goods in the source countries. And, third, price increases can overcome the productivity disadvantage. However, as a path to prosperity, Aruba's tourism development, based on the SSS model, was no guarantee it could solve the socio-economic challenges of small islands on a sustainable basis. Was the outcome of the tourism development pathway chosen by, for example, Aruba an aberration, or was it pure luck, or did tourism development involve channels that could lead small islands to economic growth and job creation?

According to the Lejarraga and Walkenhorst's study, increased tourism to small islands has drawn local businesses into interaction with the tourism industry in a manner that increases the productivity of goods and services. As tourists reveal their need for their visit to provide satisfying experiences via events and activities, creative business sense has responded with more significant and more diversified offerings to meet tourist demands. This interaction, then, propels increased product development and sales for resident businesses. As tourist demands emerge, local business owners can grasp the potential for increased profits that parallel their ability to create the goods and services that tourists require. For example, the cultural tourist may desire to experience the indigenous local cuisine or music genre. To satisfy that desire, local businesses might make entrepreneurial efforts to supply tourists with that experience. As interdependence is created between businesses to supply tourist its wants and needs, and as businesses adapt to the need to link with one another, the small island economic condition is encouraged to grow.

This more optimistic perspective regarding tourism's developmental role in small economies is validated by the evidentiary reality of several small islands in the Caribbean. For example, a study by Ramkissoon published in 2002, concluded that tourism development played an essential role in the economic performance of small islands such as the Bahamas, Barbados, and Antigua and Barbuda compared to those islands where tourism had only a modest presence in their economic structure. This study assessed the change in income per capita over 25 years from 1975 to 2000. Another study from Grassl in 2002 (as cited in Jayawardena and Ramjeesingh (2003)) found a significant correlation between tourism growth and economic growth when reviewing 29 Caribbean countries' economic performance. Other commentators also weighed in on this subject and concluded that tourism development works well economically in small islands.[20]

Recent global empirical evidence showed that small islands were ahead of others in per capita GDP. Compared to many developing countries, small islands exhibit an enviable record of economic performance. This income advantage largely reflects a productivity advantage. It is evidence either against the inability of small islands to exploit increasing returns to scale in the wake of global markets or for the irrelevance of this criterion. Evidence suggests these countries are wealthier and have higher productivity levels than large states, and "grow no more slowly than large states." Country size variables have virtually no influence on differences in growth rates among countries.

These patterns are confirmed with Latin America and the Caribbean. Countries with fewer than 1 million inhabitants have outperformed larger countries with more than 10 million inhabitants by a ratio of more than five times during a period from 1980 to 2000. The largeness of scale is no guarantee of prosperity. Conversely, the smallness of scale is not fatal to prosperity. In comparing performance in tourism value creation of six regions in the world with the Caribbean region, I found that the latter has outperformed the other regions in generating more value-added in the supply of tourism products over a period from 1986 to 2001. Prasad pointed out how Mauritius was able to build an enviable manufacturing industry despite being a small island. Antigua and Barbuda and Barbados were able to use the services industry to their advantage and propelled economic growth. This empirical record suggests that either the constraints alleged by small size are not so severe, or these small entities overcame these constraints.[21]

This finding questions the validity of the assumptions of the neoclassical view of economics and development. This is a comforting and encouraging conclusion because it shows that success is possible despite the small size. A change of perspective from size to policy, however, is like a paradigm shift. In Kuhn's classic study of the role of paradigms in science, it is the

---

[20] See, for example, Seetanah (2011); and Algieri, Aquino, and Succurro (2018).
[21] I will update these comparative patterns in Chapter 5.

accumulation of repeated empirical rejections ("anomalies") of the traditional theory that warrants a fundamental modification (paradigm shift) in scientific thinking. Scholars of small islands now have many anomalies that cannot be explained by small states' mainstream economic theory.

Was this extraordinary performance the result of pure luck, or was the chosen pathway the right one for small island destinations? Is TS inherently suitable for small island destinations to overcome the barrier of size, and can small island destinations realistically pursue a growth economy? Moreover, if TS is a realistic pathway to growth and prosperity, how should small island destinations pursue this specialization strategy?

While tourism from a global perspective appears to bear out positive growth straits, the same is not necessarily true of small island destinations. That is, tourism growth at these destinations appears to be characterized by a more erratic or unbalanced nature. The possibility exists that tourism may be unduly relied upon. Government officials, investors, and travel agents may become so enchanted with tourism that they exaggerate the total market size or ignore its competitive nature. Even though they may favor increasing the allocation of financial resources to the tourist sector to increase income levels, policymakers, and others typically do so without the support of empirical analyzes. In either case, the tourism sector will not achieve its full potential and will not provide its optimal development impact. Thus, any tenacious relationship between tourism volume and economic input would be challenging to grasp resolutely. Some studies apply computable general equilibrium data to reveal the possibility that tourism development and its growth could change or replace other economically viable components of the destination's economic development, thus limiting overall economic gains that might otherwise be attained. [22] On the other hand, some studies reveal both fast and efficient growth in the long-term.[23]

3.4   TOURISM SPECIALIZATION (TS) AS A PATHWAY

TS seems the pathway that served as a lifeline to the Caribbean countries and other small islands such as the Balearic Islands and the Canary Islands.

---

[22] See, for example, Adams and Parmenter (1992); Dwyer and Forsyth (1998); Dwyer, Forsyth, Madden, and Spur (2000).

[23] See, for example, Shan and Wilson (2001); Balaguer and Cantavella-Jorda (2002); Vanegas and Croes (2003); Brau et al. (2003); Durbarry (2004); Dritsakis (2004); Eugenio-Martin et al. (2004); Neves-Sequiera and Campos (2005); Kim et al. (2006); Croes and Vanegas (2008).

Specialization patterns have played an essential role in economic performance of these islands. While extractive and goods-producing economies have shown more significant difficulties of adaptation, service-based economies (such as Barbados, Antigua, and Barbuda, St. Lucia, St. Kitts and Nevis, St. Vincent and the Grenadines, the Bahamas, the British Virgin Islands, the Cayman Islands, Malta, the Balearic Islands, the Canary Islands and Mauritius) have shown a more robust response to globalization.[24] Their trade and growth performances have surpassed those of goods-producing economies, except, for example, Trinidad, and Tobago. Usually, for small-yet fast-growing-countries, the predominant productive sector has not been oil extraction or manufacturing but has instead been the service sector, including tourism.[25]

The lackluster performance of the conventional industries and the economic prowess of TS prompted international organizations to revisit their view on tourism.[26] As such, the Task Force on Caribbean Reconstruction Facility, a joint committee of the Caribbean Development Bank and the Inter-American Development Bank, recommended service exports (tourism) as a strategy for enhancing the competitiveness of Caribbean economies. However, such a recommendation runs counter to several long-held notions about the limits of tourism as an engine for economic development. The tourism literature lodges four main critiques to the tourism development model. First, tourism is incapable of generating any significant productivity gains, so how could countries specializing in tourism hope to achieve high income and growth levels? Heavy reliance on tourism has been claimed to have adverse effects on growth and development, mainly through high leakages for imports and displacement effects on other sectors. Specialization involves high opportunity costs of resource allocation and has therefore been dismissed as a development strategy for small countries.[27] Second, as international demands can be given to fluctuations, the smaller economy could fail to meet new demands as their product inventory leans heavily toward specialization. This specialization can act as a deterrent to meet the requirement of scale; thus, there emerges the need to import the goods and services that might otherwise work to produce for domestic consumption while exporting their

---

[24] On the effects of globalization, see, for example, Read (2004).
[25] See, for example Ocampo (2002); Armstrong and Read (2000).
[26] For an excellent synthesis about the tourism literature on small islands, see the work of Parra-Lopez and Martinez-Gonzales (2018).
[27] See, for example, Wilkinson (1987).

goods and services to augment the stagnant scale. The distressing danger is that small islands may risk becoming victims of the global market if they continue the export process.[28]

Third, small islands seem to have less ability to deflect volatility. The result of this liability viewpoint is that it portrays the small economy country as hopelessly insolvent, and its prospects for sustained prosperity are desperately impossible. Moreover, the fruitless economies of small islands indeed are characterized as beggars and takers requiring financial aid. This hopelessness has not fallen on deaf ears. Some have taken up their cause and advocated for help to compensate for the small islands' exposure to global victimization.[29] And yet, small island destinations seem to have overcome the island paradox by engaging in trade and sidestepping scale. And fourth, market location and size are factors impacting business costs. As small countries are challenged with limited access to resources and may lack strategic development plans that could lead to sustained growth by their abilities to secure and sustain their economic future, they instead are often consigned to service roles. This service role could establish a cycle, at which point foreign service companies move in and employ local residents in lower-skilled jobs. Generally, these jobs provide limited professional growth opportunities for locals. Thus, the local workforce could become inept in accessing its own creative agency that would propel economic development. As a result, the revenues earned by foreign service companies filter to their coffers while paying far less to the local workforce.

In his study, Smeral found that tourism services have become relatively more expensive than other goods.[30] He attributed this trend to the unique nature of tourism production and consumption. First, tourism reveals a perishable production process (i.e., an unsold hotel room is lost forever). Second, there is an element of inseparability in production. There is no special separation between production and consumption, as the customer must be on site. The coincidence of production and consumption both temporarily and spatially restricts rationalization opportunities in the tourism industry. In Smeral's view, because productivity in the tourism sector job lags behind productivity in manufacturing, the tourism business's costs rise. The underlying assumption in his reasoning is that nominal

---

[28] See, for example, Briguglio (1995, 1998); Easterly and Kraay (2000); and the special editions of World Development, 8(12), 1980 and 21(2), 1993.

[29] See, for example, Briguglio et al. (2005).

[30] See Smeral (2003).

wages equalize across sectors, so lower productivity shows up in relatively faster increases in prices.

However, a paradox seems to exist when examining small island destinations in that they seem to have overcome some of the size constraints of being small by sidestepping scale and engaging in trade. There are still a large number of tourists every year. International tourist arrivals had increased from 25 million in 1950 to 528 million in 1995 to 1.4 billion arrivals in 2018.[31] This uninterrupted growth suggests that, while the cost is essential, it is still not everything. Instead, people want to enjoy a unique and memorable tourist experience. Therefore, lower productivity alone cannot explain the steady increase in tourism prices. Demand for tourism is a moving target. As people become more prosperous, they typically want higher quality products. Creating these quality products takes more time, effort, energy, and talent. This process is what we know in economics as unbalanced growth; resources are shifting to the economy's low productivity growth.[32] In a study that I published together with Ridderstaat and van Niekerk, we noticed in Malta that human capital was critical in delivering quality offerings and services to tourists, more so than capital investments. I have more on this unbalanced growth in Chapter 7.

3.5   REKINDLING THE TOURISM SPECIALIZATION (TS) MODEL

What does all this mean for TS? The comparatively inelastic reaction of supply to rapidly changing demand affects the adoption of productivity advances and innovative potential in the tourism industry. Tourism products can extract higher 'rents' because of scarcity. Scarcity is characterized by small size and isolation, as suggested by Scheyvens and Momsen in their study of 2008. These features are valuable in TS due to their appeal to international tourists. These authors assert, "Isolation is often considered a drawback to those trading products around the globe, but for tourism, it may be a benefit in that it tends to make the destination more attractive and exotic, especially in the case of small islands. The demand for holidays to small island states is growing… islands are the second most important holiday destination after the category of historic cities…" (p. 498).

---

[31] See UNWTO (2019).
[32] We observe this tendency in Western capitalism, where the bulk of employment moved away from manufacturing to the service sector.

Moreover, tourism is a luxury good. As a luxury good, tourism goods tend to compete less with the demand for other non-tourism goods as income increases. The combination of scarcity and income elasticity can overcome the tourism productivity lag and lower the opportunity cost of TS.[33] This combination implies that as income increases, consumers in high-income countries tend to spend a higher proportion of their income on vacationing while only slightly influenced by price variations. Attaining cost advantage in tourism production is challenging, if not impossible. The only way to gain cost advantage seems to be on the demand side by making the tourism product more appealing, subsequently extracting higher prices over time.

To augment and strengthen the tourism industry, the entire destination is immersed in developing and altering its store of goods and services. Tourists' wants and needs are fluid and, therefore, destination offerings must shift, change, or progress to provide for tourist demands. These demands filter to related tourism industries such as transportation, event planning, and lodging. Moreover, as tourists seek to experience small island goods and services, the available supply of local products must meet their demand. Potential for costs and profits are then encouraged by the tourists' demands and availability of supply.

But higher prices for goods and services could also be attributed to the particular consumption pattern of tourism services. There is growing evidence that tourism is a luxury good because tourism demand has an income elasticity above 1.[34] This implies that tourism services demand is triggered disproportionately in an individual's budget when incomes rise. As tourism products compete less with non-tourism goods over limited budgets, the more price increases can overcome the productivity disadvantage.

A price policy, extracting a "rent" to offset the difference due to the productivity lag, will work best however, in a high quality, unique tourism-based environment. Keane (1996) posits that quality revealed in the product and delivery guarantees optimizing the economic performance of the tourism sector. Under such a condition, can a small country make up for this deficit by facing lower opportunity costs of specialization? Overcoming the productivity disadvantage provides benefits. Two relevant

---

[33] However, there is a caveat in this thought process as the income impact on tourism demand may be changing due to the effects of, among others, business cycles and changing tastes and preferences.

[34] See, for example, Croes (2010). Croes and Vanegas (2005); and Vanegas and Croes (2003).

issues about these generated benefits are worth mentioning. The first issue refers to who will appropriate these benefits. Will increased prices induced by scarcity go into private individual's pockets rather than to the community to which it belongs?[35] The second issue refers to the guarantee that the person or firm can make the best use of these benefits. Combined, these two issues frame whether tourism benefits contribute to the destination's well-being.

The tourism product, experience, is collectively created by the destination. This experience aims to satisfy the wants of the tourist. The tourist engages in a decision-making process involving time allocation (work, leisure, or vacation), while the supplier engages in resource allocation (staffing, procurement, and amenities). As resources tend to be scarce, choice decisions regarding resource allocation must be made. All these choices are time and place-bound and are related to the enjoyment of a place. Unlike mainstream economics, where supply and demand meet in markets and location seems irrelevant, space plays a fundamental role in the tourism activity. This fundamental role is reflected in the measurement of tourism flows referenced in a specific spatial definition of a destination (e.g., arrivals, length of stay, spending, carrying capacity).

Pertinent to the choices is the consideration that tourists demand unique and satisfying experiences, the success of which falls mainly upon the locals. That is, residents can posture in a positive, hostile, or indifferent manner toward the tourists, defining the effectiveness of the goods and services to comply with tourist expectations of the unique experience they seek. Thus, the combination of offerings and local interaction merge to satisfy (or not satisfy) tourist wants and needs. This merging could be the bedrock that creates the possibility for tourists to adapt their tourist experience according to those attributes they define as favorable to their quality of life.[36] While small island destinations grasp the need to strategize by way of offering the types of goods, services, and unique experiences for tourists, there is a trade-off that could compromise the welfare of the destination and its people and the small island's opportunity to move toward prosperity. That is, as large-scale development occurs, the treasured and limited resources that may have attracted tourists in the first place, and that flourish only in a

---

[35] As I stated in chapter two, tourism specialization is a collective process.
[36] See, for example, Richards (1999); and LaSalle and Britton (1982).

delicate balance of nature may be compromised, abused, or destroyed by insensitive or unknowing tourists. Such a loss is devastating-breeding bitter antipathy in the locals.

Because tourists evaluate their experiences according to destination offerings and interactions, then destinations are compelled to strategize their decisions and choices to provide tourist opportunities in the most advantageous manner possible. The result brings lucrative prospects for the destination to attract and capture new and repeat visitors while competitively dominating price and profit. Understanding the tourism market and its relationship with the destination as the unit of analysis is essential for resource allocation involving marketing, infrastructural investments, and required coordination for optimal economic results, and environmental preservation. Optimal economic results are captured through receipts and their impact on economic growth. Despite the great interest in international tourism and the attractive estimated growth in tourism receipts, there is a danger that Caribbean countries may ignore or squander their tourist potential because of a fear of the high infrastructure costs involved in tourism development. Despite the great interest in international tourism demand and the growth in receipts, optimization of the Caribbean Basin's tourism industry is a pressing matter. The region is losing market share, and the expenditure per visitor appears to decline over time (WTO, 2005).

These islands may lack knowledge of the market demand for tourist facilities. They may fail to grasp the extent to which economic growth responds favorably to the buoyant world demand for tourism. This lack of knowledge has created fear in tourism prowess as a development strategy. This fear is fueled by the claim that heavy reliance on tourism may negatively affect growth and development, mainly through high leakages for imports and displacement effects on other economic sectors.

A profit optimizing strategy could have an additional benefit for the destination because high prices may be interpreted as signals of high quality. Signals play an important role in tourists' behavior. Many things we want to know about the tourists are not perceivable such as loyalty and emotional states (are you happy or satisfied?). We rely on signals to decipher these qualities. However, in a competitive setting, signals are difficult to decipher and can be deceptive (cheating). Veblen chronicled the wealthy's conspicuous behavior, suggesting that it is not enough to be wealthy, but there is the requirement to display wealth.[37] However, for leisure to convey prestige, status, or power, it must be wasteful. Waste is perceived as excess costs beyond what centers this behavior in the realm of money and time. Because we pay attention to how and what others conclude from our choices, tourism has the unique opportunity to spawn excessive spending and convert destination experiences into luxuries.

Tourism as an experience responds to tourists' wants. Empirical evidence suggests that people derive longer-lasting satisfaction from spending on experiences than on material goods.[38] Demand moves quicker for wants, so if you can satisfy wants you probably have a tourism product upon which people will spend proportionally more of their income than spending it on needs. Demand will rise faster as income grows. As a luxury, good tourism is income elastic, so demand increases proportionally faster than income.

Due to intangibility, one buys the tourism product (a destination's experience) before one can verify its quality. This means that one must rely on the reputation or image mechanism to cope with this uncertainty. According to Keane[39], charging premium prices is an incentive to deter the tourist industry's components from cheating on tourist product quality. He identified Bermuda as a high-quality destination that charges premium prices to undertake quality maintenance of its hotel inventory and increase its reputation to attract more customers. Reputation is an essential feature under conditions of imperfect information, where tourists visiting a destination have inadequate information regarding the reliability and preferences of vendors or suppliers of goods and services at the destination. Essentially, this is a question of how much trust a tourist places in others at a destination.

Suppose a destination can charge higher prices through reputation. In that case, the costs of foregoing growth effects from knowledge accumulation in manufacturing and other "progressive sectors" (according to the endogenous growth theory) may be outweighed by trade through importing productivity growth overseas. Because the tourism product is income elastic, residents from destinations and tourism source countries can benefit simultaneously. Destinations benefit from ToT gains induced by tourism. At the same time, tourists enjoy memorable experiences during their vacation-absorbing mores of other cultures, imbibing foreign cuisine,

---

[37] See Veblen (1994).
[38] See, for example, Pine and Gilmore (1998).
[39] See, Keane (1997).

socializing with locals, participating in events and activities, and visiting historical sites. Overall, then, vacationing can be important in that the memorable experiences it provides can provoke new, useful, and valuable perspectives and ideas that are advantageous to increase the tourists' quality of life. Growth patterns spurred by TS have worked as a lifeline in the Caribbean when confronted with difficult policy choices. The record in the Caribbean and Aruba bears out a competitive advantage in tourism.

And yet, the most effective means for achieving economic performance on a small island continues to be the subject of considerable study and debate. Theories abound on how to trigger economic growth ranging from comparative and competitive advantage, the openness of an economy, and economic scale and scope hurdles. Adam Smith, in his seminal work The Wealth of Nations, published in 1776, argues, contrary to his predecessors, that economic growth is founded on productivity and productivity is turned on by specialization. His specialization theory asserts that by dividing labor into smaller and transparent tasks, economic activity would expand, create a market between other countries, and spur economic growth.

I will assess and discuss this claim and the relevance of demand in more detail in the next chapter.

KEYWORDS

- **business organizations**
- **globalization**
- **gross domestic product**
- **industrialization**
- **tourism development**
- **tourism specialization**

CHAPTER 4

TOURISM ANGST AND DEMAND PUSH

As I delved deeper into the literature that addresses small islands economic prospects, I was struck by the scant research available about why and how small islands can grow and overcome their seemingly inherent constraints. The literature has been primarily concerned about an island's features as a geographically bounded place, emphasizing smallness, vulnerability, and relative insignificance. Small islands were confronted at the end of the 20th century with a rapidly changing global landscape, threatening the extant small island developing model's core. This model was based on import substitution and trade preferences. Switching to tourism development was directed more as a guts stroke than decisions based on a deliberate data-driven choice. Regrettably, small islands are at a crossroads, straining to immunize themselves against the international system's uncertainties.

My multiple conversations with leaders of small islands in the Caribbean corroborate this depressing view. While these small islands were desperately searching for growth to subdue their economic and social problems, they seemed guarded to embrace tourism. There appeared a lack of understanding and tenuous enthusiasm regarding tourism development's prowess and its potential developmental role for small islands. This hesitancy seems justified when considering that empirical research has indicated that tourism development benefits may be evident only in the short-term. From a long-run perspective, it may be that the benefits are unclear at best and, at worst, may be too risky to warrant enthusiasm. Indeed, this is understandable as its non-linear properties characterize tourism development, thus revealing outlooks of uncertainty.

I wanted to answer the question related to whether tourism specialization (TS) could be the right choice for small islands. The words right choice was assessed from an economic growth perspective. Development economists insist that economic growth is vital for three reasons. Firstly, these economists believe that when we produce more, we earn more income. More income has three effects according to this perspective: we can consume more, we can increase the standard of living, and we can enjoy a happier life.[1] Secondly, we need to produce more to keep up with population growth. For example, this perspective reasons that if the average annual population growth rate of Mauritius from 1960–2018 is 0.06%, then the annual economic growth should be at least 0.06% to merely remain static. Otherwise, unemployment would rise, wages would fall, and the standard of living would decrease. And thirdly, this perspective asserts that economic growth is the only way to reduce poverty. If the economic pie does not grow to include more people in its advantages, the result could be political instability and rancorous under-development.[2]

According to mainstream economic literature, the question following these economic assertions is, can tourism induce economic growth in a small island despite its seeming impossibility? This question brings me to specialization as discussed in Chapter 2. Since the time of Adam Smith, it has been asserted that specializing is defined by how cost-effective you can be in generating a product and whether you have the available resources to make such a product. Smith's cumulative model of growth is anchored on the ability to specialize, which depends on scale (market size) and the division of labor. Smith's conception indicates that the division of labor is associated with productivity, and the latter is related to income. This full growth circle implies that the structure of production matters for economic performance because the economic sector with the most promising growth potential should be pursued and supported.

While each country, unless a colony, has the freedom to decide what to produce, effective freedom is constrained by the country's opportunities and abilities to produce. A large country may have a broader array of economic opportunities than a small island whose array of opportunities may be small. Alternatively, a country with a valuable resource such as oil may have more economic opportunities than an oil poor country. An oil-rich and large but poorly organized country may have a smaller scope of opportunities compared to an oil poor and small but well-organized island. Countries must choose what to produce, and actual and potential opportunities determine this choice. Opportunities matter and choices are consequential! Moreover, this is precisely the challenge that small islands

---

[1] For a recent account of this perspective, see, for example, Stiglitz, Fitoussi, and Durand (2019).
[2] See, for example, Easterly (2001).

face due to their small scope of opportunities. It is clear, scale defines the level and quality of opportunities-exactly the challenge for small islands. The abilities to effectuate products based on its opportunities are a formidable issue for small islands.

Indeed, the literature is full of disadvantages that small size imposes on growth processes, rendered by the difficulty of achieving sufficient economies of scale in a wide variety of basic economic activities. Because of the built-in restrictions related to market size, resources, labor, and capital, small countries are forced to look to the global economy to achieve economies of scale in the production of exports. Increasing exports means expanding the size of the domestic market. However, the island's small geographic size is often used to characterize the potential of the island as too diminutive to harbor a successful piece of the market via expansion of its domestic market for export purposes. For example, William Demas has written about the Caribbean islands: "Small may be beautiful; but it may also be fragile, vulnerable, and extremely externally dependent." Small islands simply lacked the resources and were subject to high transportation costs for importing and exporting raw materials and finished goods.[3]

Demas's view may have been shaped by the realities on the ground. The development model practiced by small islands burgeoned after World War II. This model, based on agriculture and manufacturing, revealed halting success, and came to a screeching halt in the 90s. For example, the "industrialization by invitation," propounded by Arthur Lewis, an economic Nobel Prize winner, had run its course because small islands were ill suited for manufacturing goods. The Lewis model was suggested to answer the lack of economic growth that the Caribbean region was experiencing after World War II. The anemic economic growth was due to extensive underemployment triggered by the inefficient agricultural sector. This inefficiency results in more workers in the agricultural sector than were required. Earning subsistence wages hampered the worker's exploitable savings that could otherwise have been used to provoke investment for growth.

Also, small islands further suffered a decline in their economic proficiency and stability as global trade preferences for small island products could not meet the demands of growth required as an economically sustaining factor. A somewhat domino effect would then ensue in that the small islands could

---

[3] The Journal of Development Studies dedicated a special issue on the challenges facing small island developing states in its edition of 46(5), 2010. See also, for example, Briguglio (1995).

not meet the rigorous demands of competitiveness in a global market. As demand for the more traditional small island products waned, struggles to recoup often resulted in poor management of resources that bolstered additional economic issues.

Also, infringing on small islands in the Caribbean is the specter at the end of Cuba's embargo. Travel restriction to Cuba from the United States had been imposed since 1963 due to the Cuban Revolution in 1959. Once the U.S. denied Americans access to Cuban tourism opportunities, other small island prospects to vie for U.S. tourists emerged. This opportunity meant that small islands were no longer insulated from the stark competitive offensive into which they were thrust post-Cuban restrictions. What became evident was the possibility of capturing a market share of tourists. Cuba's absence from the U.S. to the Caribbean market afforded a competitive circumstance that appeared advantageous to small island prospects.

However, the risk could materialize that could be hazardous to small island destinations should U.S. tourist travel to Cuba reopen. That is, questions such as whether Cuba would regain its past prominence in attracting U.S. tourists would loom, and the status and stature of the Caribbean small island destination might be uncertain. This means that the Caribbean tourism community could undergo the types of considerable modifications that could alter the tourism landscape in manners that may or may not impose difficult conditions for the small island destination. Arguably, this easing of constraints could have a ripple effect on those islands close to Cuba. We discuss the potential dangers of Cuba's reopen to U.S. travel later in the chapter.

4.1 A CREATIVE AGENCY PATHWAY TO OVERCOME STRUCTURAL CONSTRAINTS

Easterly and Kraay argued in their study published in 2000, that service specialization has a strong positive impact upon economic performance. Export and trade are growth enhancing due to their focus on growing productive capacity through foreign markets. Openness through trade is based on a comparative advantage framework; it triggers domestic competition, which hampers rent seeking (political economy benefits) and encourages faster productivity growth (growth benefits of trade). Trade by way of supporting export growth provokes growth in all other demand components by paying to import, invest, government expenditures, and exports themselves. Also, export growth pays for the import of many goods to facilitate economic development. In other words, export growth has demand and supply-side effects in the economy.[4] The implication is that ToT variations can explain growth differentials performance among countries. Moreover, the connection with the outside world is the channel through which small islands can grow.

In 2002, Balaguer and Cantavella-Jorda were among the first commentators to demonstrate the relevance of tourism on long-term Spanish economic growth by revealing how foreign exchange earned through tourism paid for Spanish imports. This study opened the door for tourism as a long-term economic growth pathway. Developing tourism in these small countries (e.g., more hotel rooms and more arrivals) is one plausible explanation for why their economic performance in growth has not been diminished beyond its current state. As a service, tourism is a subjective experience consisting of multiple services and goods spread geographically over a destination. If tourism as an experience can be sold to international travelers, tourism can be an export-led strategy based on services.

Nevertheless, the potential for tourism as a developmental tool is still in discussion, mainly since development results based upon tourism are still inconclusive. Despite tourism's growing importance as an engine of growth for many countries, it has received little attention in economic development literature. Not everybody is convinced of tourism's prowess to propel growth in small islands. For example, William Demas wrote in 1965: "It is true, tourism is highly income elastic, but it depends so largely on whim and fashion that it would not be prudent in countries where it is possible to develop manufactures to place hope entirely or largely on this industry."[5] Negative sentiment against tourism was again echoed 25 years later by the study of Pastor and Fletcher. They cited tourism's contribution of much needed foreign exchange to Caribbean countries as a weakness rather than strength. They considered tourism to be an unreliable source of revenue too easily influenced by exogenous factors.

Political economy frameworks also prompted a critical view of tourism opportunities and challenged the optimistic views regarding tourism anchored in the modernization frameworks. The study of Emmanuel de Kadt expressed this critical view. This study, published under the World

---

[4] See, for example, Setterfield (2002).
[5] See Demas (1965).

Bank's auspices, was critical of tourism's role in developing countries to spread the benefits to those in need and stressed the social, cultural, and environmental costs. Some commentators assert that tourism is a form of neo-colonialism and that tourism prompts dependency through an "enclave model of third world tourism." His study was a precursor of applying the 'dependencia' framework to tourism analysis.[6] Other studies did not even consider tourism as a development pathway. The 2001 work of Bernal, Bryan, and Fauriol attempted to influence the US foreign policy agenda of the Bush administration. These studies seem more concerned about the decline of official development aid, and deterioration of preferential trade agreements based on subsidies and quotas for commodities such as sugar and bananas than tourism.

Qualms persist about tourism, particularly its purported weak resilience in the face of natural and manmade disasters, income changes, and the vicissitudes of fashion. There are charges that the demonstration effect can exacerbate socioeconomic imbalances in certain fragile microstate societies. These deleterious effects on the local social fabric may include the adoption of foreign lifestyles and values, the incidence of poverty, increasing crime, prostitution, and resentment within the host population, which, according to some critics, are the result of the direct interaction of tourists with the local population. According to these commentators, these dynamics are beyond the control of tourism planners and businesses of the host country. This critical attitude, although diminishing, is still prevalent today in certain quarters. Notwithstanding that prevalence there is evidence and confidence of tourisms promise for small island economic growth.

4.2   THE TOURISM EFFECT

My studies chronicled three reasons tourism works.[7] First, international tourists increase the market size mitigating domestic market constraints. Second, in competing for international tourist markets, the local market is compelled to raise and improve its service standards and efficiency. Third, providing scale and competition improves residents' quality of life. In this light, some countries have selected tourism specialization (TS) as a deliberate economic growth strategy to answer how to achieve greater economic

---

[6] See, for example, Britton (1982); Weaver (1988).
[7] See Croes (2006).

and development performance. Yet, while mainstream economics has considered several determinants of growth, including capital, land, and labor, it has failed to consider tourism's role.

However, even if socioeconomic pitfalls exist, there has been no shortage of small island interest to view tourism as a potential source to fuel future development scenarios. Tourism can contribute to establishing the road to success. It is the largest service industry in the world today and the largest single item of world trade. Further, the tourism industry has been steadily growing, despite periodic fluctuations. The World Tourism Organization (WTO) documented the results of global tourism activity in 2001, indicating that international tourist arrivals had only slipped by 0.6%, despite the tragedy's adverse ripple effects on 9\11.[8] Again, tourism has shown a great capacity to rebound after a downturn showing a V-pattern. Indeed, during times of recession, tourism is less affected by cyclical fluctuations than other economic sectors as the largest service industry.[9]

Tourism is also more resilient and less volatile than other economic sectors. The literature provides evidence to support this claim. For example, a study of tourism in the Caribbean by Maloney and Rojas concluded that tourism revenues range from two to five times more stable than other goods, including agriculture and manufacturing. This claim suggests that tourism revenues are likely to be less volatile than these other commodities. Similarly, Easterly, and Kraay contradicted the extant conventional wisdom that volatility triggered by openness due to tourism will make small islands more vulnerable than larger countries. Their study found that "...the positive growth effect of openness ... is 2.5 times larger than the negative effect of small states' greater output volatility."[10] Tourism also grows faster than other economic sectors in developing countries.[11]

Because it is subjected to constant international competition, the local tourist industry in any particular Caribbean country must improve and raise its standards. For example, the local hotel must maintain high international standards wherever it operates, and it will impose such high standards on its local suppliers and employees. In both products and services, the local market must raise and maintain its standards. The country need not worry about trade restrictions, because the customers, unlike other goods,

---

[8] Not all regions were equally affected. For example, the Caribbean dropped by 3%, while South America and North America dropped by 6.2% and 6.8%.

[9] See, for example, Kammas (1991).

[10] See Easterly and Kraay (2000).

[11] See, for example, Lanza and Pigliaru (2000).

are 'imported' into the host country. Tourism enjoys no form of protection because destinations must compete not only to survive but also to reach optimal economic output levels. They compete based on product quality and efficiency. This competition makes the domestic market work effectively and promotes efficiency in the entire economy, not only in the export sector.[12]

Beyond this, there is a deeper meaning to support TS. A country is better off with tourism due to its high-income elasticity. As long as tourism income elasticity is higher than the sectoral income elasticities of demand for imports, tourism seems growth enhancing. Demand characteristics of exports and imports define economic growth, and TS can defeat Prebisch concerns of a skewed balance of payments negatively affecting developing countries' growing opportunities, including small islands.[13] Efficiency and productivity and their association with growth may be more related to superior characteristics of demand and less with efficiency stemming from superior technical characteristics.[14]

This line of reasoning implies that superior demand characteristics embedded in tourism may overcome the superior technical characteristics anchored in the Schumpeterian growth model. The latter asserts that innovation and skills expand and transform the production potential by replacing or displacing obsolete products and firms, by higher productivity, or nurturing new niches and industries.[15] For small islands to engage in innovation as a vehicle for growth and prosperity similar to large countries is almost impossible because they lack scale. That is why small islands typically engage in Smithian growth to overcome the penalties of size by relying more on specialization gains resulting from trade.

Until now, the typical arguments favoring the allocation of more significant resources to the tourist sector to increase income levels have not been based on empirical analyses. Therefore, small island destinations continue to struggle to establish the overall level of marketing support that should be provided. This matter is pronounced in the Caribbean Basin. Already in 1996, Jean Holder, the then general secretary of the Caribbean Tourism Organization (CTO), expressed concern that few individual

---

[12] See, for example, Bhagwati and Srinivasan (1979); Krueger (1980); Holder (1996); Clancy (1998).
[13] Raul Prebisch is an Argentinian economist who recognized that unequal trade in the international system is the source of underdevelopment and has a widespread influence on its followers' dependencia framework. See Prebisch (1951).
[14] This reasoning follows the central tenets of Kaldor's growth model. See Kaldor (1985).
[15] For a discussion on the term innovation, see, for example, Fagerberg (2003).

countries had the marketing budget to compete effectively in the global market. He advocated for a regional marketing approach to create one regional brand: The Caribbean. To a certain extent, his strategy worked, building on the popularity of the *Kokomo* song by the Beach Boys at the end of the 80s. That song consecrated the Caribbean region as a favorite vacation place for the American traveler.

By the year 2015, the Caribbean earned more foreign exchange by tourism compared to Canada, Mexico, and all of South America and Central America. In receipts, tourism in the Caribbean generated the U.S. $28.5 billion in foreign exchange, representing 14% of U.S. receipts, 172% of that of Canada, 160% of that of Mexico, and 110% of South America, and 250% of that of Central America. The Caribbean region consists of one of the largest insular densities globally and depends on tourism more than any other region of the world. Tourism is the largest earner of 16 of the 29 countries, and Sun, Sand, and Sea are still an integral part of the region's charm and resource to attract tourists. One of seven jobs in the Caribbean is supported by tourism, and by 2025 tourism would generate nearly 2.8 million jobs.

Economic growth, achieved primarily through tourism, has not kept pace, however, with the labor force's increase, resulting in unemployment and underemployment, and poverty. The total population of the Caribbean is projected to increase to 42 million by 2020, raising the employment rate by 1.75% per year, reaching 5.4 million persons, according to the World Bank.[16] This socioeconomic situation represents the most critical challenge to Caribbean economies in the years to come, which together must match the world rate in productivity growth besides a minimum regional GDP growth of 1.75%. However, the Caribbean economic performance from 2013–2017, especially those of the tourism-dependent countries, achieved growth rates of 1.7% annually, which almost reached the World Bank's target set.[17]

4.3   GETTING AHEAD OF THE CURVE

The Caribbean, in general, has been practicing tourism for about 100 years. In her study of 1999, Schwartz chronicled the history of tourism

---

[16] See Vision (2020). *The World Bank Group*. http://wbln0018.worldbank.org.
[17] Visit the World Bank; link at: https://www.worldbank.org/en/country/caribbean/overview.

as it evolved in the Caribbean region. The study suggested tourism transformative potential in Cuba and the Caribbean. For example, the study narrated how Cuba became a chic resort destination for the American tourists. Cuba's tourism development bounced from boom to bust and back. This oscillation in fortunes appears to reflect the small islands' experience regarding tourism development. Buswell also recounted this experience in the case of the Balearic Islands.

Small islands seem ill prepared to face oscillations and fluctuations in the global market. Bryan asserts that the smaller islands in the Caribbean are also ill prepared to seize the opportunities that technology can provide them to market and promote themselves more efficiently.[18] Many island destinations in the Caribbean region desperately need growth. However, they do not get the maximum advantage from their tourism potential because they lack empirical knowledge of the tourism market and the factors affecting tourism demand. Instead, they rely on anecdotal and ad hoc advice in designing and implementing their tourism policy. Tourism development programs may fail because they have not evaluated these factors properly. Unfortunately, many small islands still give a low priority to tourism information based on scientific research. If the Caribbean countries, for example, continue to ignore the importance of scientific data and rigorous analysis, they will soon find themselves unable to compete effectively in tourism's global marketplace. These small islands must confront the fast-changing international tourism landscape. Developing countries are no longer passive players who see their citizens off as tourists destined for the Caribbean; they too have embraced tourism as a growth tool. Increasingly, they are competing with the Caribbean. The growing technological innovations in tourism marketing and the developing countries' inability to master such innovations more readily could jeopardize the industry's viability in the region with its already limited resources.

Economic growth hinges upon the rate of innovation, its spread, and its application. Innovation can help bring down costs through new processes, quality improvements, and increased profitability. Innovations are expected to increase consumer surplus while increasing profits for providers of goods and services. Tourism is well-equipped to spur innovation because it entails several characteristics that can influence the innovation process. Tourism is best defined on the demand side as manifesting

---

[18] See Bryan (2001). A study by Croes and Tesone showed that technology is perceived as a cost rather than an opportunity. See Croes and Tesone (2007).

heterogeneous and ephemeral industrial structures. The tourism sector is vital to many sectors of the economy. However, destinations are subject to life cycles affecting their competitive ability. Tourism policy, therefore, increasingly focuses on the promotion of innovation. However, small islands lack the institutional capabilities to keep up accurate and updated information about the destination consistently.

Despite the advantages of a relatively high development level, many years in TS, and years of competitive experience, the ability to compete on price suffers greatly from high wages and cost disadvantages. This is not only due to structural challenges induced by small size; this matter is also triggered by policy such as trade protectionism and poor business climate. However, a strategy to reduce pricing to remain competitive is dangerous, especially in a productivity-challenging environment such as small islands face. Racing to the bottom is never a winning proposition.

4.4   FLYING BLIND WITH CUBA?

An eventual opening of Cuba to American tourists should also be a matter of concern. Cuba may be a threat to the other countries of the Caribbean because the destination could regain its dominant position in the Caribbean. Cuba's potential as a tourist destination has often been exalted due to its history, heritage, culture, nature, and beaches. Removing travel restrictions to Cuba would likely increase the total tourist demand for Cuba.

Cuba has given tourism a vital role in the development strategy having implemented tourism in the early 1990s. Today, tourism is the number one export of Cuba, surpassing both sugar and hard currency remittances, reaching 1.8 million visitors, 1.9 billion in gross income with 37,000 hotel rooms in 2001.[19] On top of the challenges created by globalization's rapid process is the specter of the impact for the individual countries in the region vis-à-vis the eventual reintegration of Cuba in the North American orbit. Several studies allege that Cuba's presence in the USA market would negatively affect several Caribbean islands' tourism prospects.[20] By 2010, Cuba had been expecting 7 million tourists, which would have required 185,500 rooms. With such a large potential rival for U.S. tourism,

---

[19] By 1999, one out of seven hotel rooms in the Caribbean were already located in Cuba. See, for example, Brundenius (2002).

[20] See, for example, Romeu (2008); Padilla and McElroy (2007).

it would behoove small destination countries to empirically assess this possibility and plan to circumvent the impact such competition could have on its demand market.

In my *Anatomy of international tourism demand*, I warned about the potential perils of Cuba's reopening to American travelers. While President Obama opened the prospect of normal relations with Cuba, there is uncertainty about the prospect of normalization with President Trump's policy towards the island. Nevertheless, regardless of Cuba's current uncertainty, it is critical as an illustration to ask the question: what if relations with Cuba are normalized and the trade embargo is lifted? Will American tourists substitute the other Caribbean islands with Cuba? Will Cuba be restored to its formal paradise glory?

Expanding the research agenda to include one non-economic event of great interest is thus the impact of the eventual reintegration of Cuba in the North American economic order. This potential reintegration includes the further development of tourism in the region and its impact on individual countries' economies. For example, the critical question is, for which countries will Cuba's re-entry become a threat? There also are empirical questions that need immediate attention, such as estimating the elasticity of substitution among major tourist destinations within the region.

As the U.S. lift on tourism to Cuba reintroduces competition for U.S. tourists, there would follow a need for the Caribbean to reassess and update how it constructs its tourism industry should that competition present deleterious impacts on the Caribbean tourism model. Empirical data on such potential impacts would require the Caribbean industry to structure responding strategies to account for the competition successfully. Currently, few studies regarding this issue are available.

As Cuba re-establishes a U.S. tourism market share, Cuba would impose price-changing tactics to accommodate the increased travel. Thus, tourism impacts to the Caribbean countries would be impending by way of cross-price elasticity. To distinguish the cross-price elasticity of the U.S. demand for various Caribbean countries such as Aruba, Barbados, the Bahamas, Jamaica, and the Dominican Republic, I employed estimations to account for Cuba's price changes and the dynamic bearing that they might sway to the Caribbean countries. The model I used for the estimations is provided in the footnote below. There have been comprehensive and thorough studies regarding tourism flow and the characteristic demand-side elements that shape its course. These studies have been completed on a global level and pertained uniquely to the Caribbean.

For estimation purposes, I sourced annual data for the years 2000–2010, provided their calculations, accessed the Federal Reserve Economic Data Bank (St. Louis FED), and The Organization for Economic Co-operation and Development. Also, information from Aruba, Barbados, and Dominican Republic central banks and the Bahaman and Jamaican tourism boards was obtained. Several techniques were utilized, such as unit root tests (ADF, P.P., a two-stage least squares (2SLS) method, and KPSS, as well as tests displaying endogeneity, model under-identification, heteroscedasticity, and serial correlation.

The analysis of the dynamics that the Cuban prices would prospectively impose would depend on which Caribbean Island is being referenced: and would depend on whether the impact involves the long or the short run. For example, neither Jamaica nor the Bahamas would experience deleterious effects; and Barbados would indicate a long-term positive outcome. However, unlike Barbados, Aruba would suffer adverse outcomes in the long-term. Yet, even under this condition, the pricing effects would be insignificant. That is, they would be 0.0188 (below 1/50th of a percent). In spite of this insignificance, there would exist a nearly 1% effect on the U.S. gross domestic product (GDP) per capita. When considering cross-price elasticity (substitution) for travel from the U.S. to Aruba, and the impact of Cuban pricing, Aruban pricing would need to rise no less than 7% when compared to that of the U.S. Also, the pricing impact from Cuba to the U.S. should move to lower than 7%. Conjoined, these two impacts would account for the 1% increase in Aruba's tourist arrivals as provoked by the U.S. GDP per capita increase of 1%.

Also, observe that the applied dummy variable wholly includes potential cyclical recessionary effects. Therefore, it is doubtful that U.S. travel to Cuba would create substitution effects given Aruba's long run high-income elasticity.[21] Table 4.1 reveals the results.

4.5 UNDERSTANDING AND EMBRACING TOURISM DEMAND ELASTICITY

Understanding how tourists make destination choices is critical to effective tourism marketing, policy, and product development (new offerings).

---

[21] One plausible reason why the Aruba tourism industry is resilient against recessionary effects is its large timeshare inventory. The timeshare inventory involves more than 40% of the lodging sector.

TABLE 4.1 Cuba Regression Results

	Per Capita GDP	P rel AUA-USA	P rel BAH-USA	P rel BAR-USA	P rel JAM-USA	P rel CUB-USA	Dummy	Kleibergen-Paap rk LM statistic (x2) (Under-identification)	P-value	Hansen J statistic (Over-identification)	P-value	Endogeneity test	P-value
Aruba													
Long-term	0.9254***	−0.1165***				−0.0188***	0.1331**	6.3890	0.0410	0.6620	0.4158	3.6120	0.3065
Short-term	0.1042	−0.1987				0.2008	−0.5069	6.3200	0.0970	2.4190	0.2983	0.7710	0.8564
Bahamas													
Long-term	0.0923		−1.2830***			−0.0182	0.4207**	5.5810	0.061	0.3960	0.5294	5.1610	0.1604
Short-term	0.3008		−0.6506***			0.0685	0.3366	6.4570	0.091	3.4080	0.1820	3.8270	0.2807
Barbados													
Long-term	0.8454***			−0.1133		0.0010	0.2013	6.1740	0.046	2.2520	0.1334	3.9250	0.2696
Short-term	1.6191***			−0.2430		0.8055***	0.6045	7.2100	0.066	1.7580	0.4351	5.9070	0.1162
Jamaica													
Long-term	0.2451***				0.8453***	−0.0017	−0.0904**	5.4510	0.066	0.8860	0.3465	3.9970	0.2618
Short-term	1.0303				0.6424	−0.2652	0.5420	6.3150	0.097	1.0370	0.9817	5.6640	0.1292

Note: ***(1%); **(5%); and *(10%) signification levels.

As demand fluctuates following trends, consumer's expendable finances, and destinations' abilities to offer satisfying experiences to the tourist, it behooves destination management to understand the considerations that consumers ponder when choosing a destination. Questions such as: can the destination offer the tourist the kinds of goods and services that consumers demand and are attractive enough to drive visits and repeat visits? This understanding is vital if the destination is to effectively develop and market the products that meet tourists' demands, while at the same time configuring and remaining abreast of the likelihood for future demand.

Moreover, analyses regarding tourism's futures are comprised of a complex consortium of contributors (government, businesses, resident welfare, and quality of life, investors, environmental concerns, resource allocation). Attention to policies, methods, approaches, and strategies to meet the rigors of ensuring that demands are understood and met are critical and challenging given the complexities involved in coordinating and synthesizing the contributing parties. Thus, accessing qualitative and quantitative research into market demand can help plan for a destination's current and future demand and be crucial to the tourist industry's success.

Pre-emptive knowledge of the rise and fall of tourist demands could allow a destination to keep loyal visitors and capture new market segments by their expressed demand. This knowledge contributes to the efficiency with which a destination could service and sustain its clientele. In effect, a destination becomes a proactive entity-constantly anticipatory and tactically adaptive to industry demand.

Most studies of tourism applied the economic utility theory as the theoretical foundation to explain destination choice. Economic utility theory postulates that a tourist will choose an alternative destination that provides him the highest utility-which depends on a set of attributes carefully weighed by the tourist. This decision-making process underlines the condition that human behavior is imminently probabilistic, which means that it is not possible to determine a priori which destination a tourist will select. However, it is possible to estimate the factors that determine the destination choice.

The destination choice is a reflection of a set of preferences that determines either substitutability or complementarity among destinations. These preferences respond similarly to tourism-related consumption, attributes, and geographical proximity. Some studies assert that utility is derived from the time spent at a destination resulting from various attributes such as climate, beautiful scenery, activities, and other socio-economic considerations. Based on the consumer theory of choice, several demand characteristics are derived to determine and explain tourism demand to a destination. For example, the literature identifies price, income, exchange rate, transportation cost, marketing cost, and time as factors that shape destination choice.

Due to its demand characteristics, I contend that tourism is better suited to propel growth in small island destinations. Tourism is highly income elastic and can induce growth-stimulating other sectors within an economy. Buswell (2011), in his study Mallorca and Tourism, asserts that the success of Mallorca's tourism "…might be drawn from a study of 'demand' rather than 'supply' (p. 10). Mallorca is one of the first Mediterranean tourism destinations and is considered the pioneer of mass tourism. His study chronicles how demand characteristics rather than supply features have defined tourism development in Mallorca and the Balearic Islands.[22]

Demand is the fundamental gauge of the attractiveness of a tourists' destination. Tourism demand is manifested in the destination where it meets the tourism supply. The interfacing of demand and supply at the destination, revealing the complex interaction between complementary and substitution goods and services, makes the destination the unit of analysis. Tourism demand is essential because of the perishable nature of the tourism product. Overall, understanding demand requires a grasp of what comprises demand, the factors affecting the demand level, and insights of how future demand can be identified and estimated.

These estimates are essential to anyone planning future tourism developments, be it the government official engaged in maintaining the adequate level and sound quality of infrastructure, or the private investor in search of a business proposition. Tourism demand research is also invaluable for marketing and promotional campaigns by keeping a destination or business in touch with its markets. It can lead to discovering new markets, new products, and more efficient uses of products. Finally, by studying and forecasting the market and developmental trends, tourism research can reduce the risk of unanticipated changes at the destination by probable scenarios and alternative strategies.

The first requirement for any successful player in the tourism market, therefore, is research into demand. Today, and in the likely future,

---

[22] Buswell describes Mallorca as "quintessentially the creature of the tourism industry" (p. 82).

knowledge of how destinations' demand is secured and shaped requires a substantial grasp of its market base and its ongoing wants and needs. Also, the fluctuating trends of the destinations' existing market share and the potential market share amid global competition are indispensable. Without knowledge of essential information such as allocating and protecting its resources, adjust, and adapt its governance, plan infrastructure, create products, and plan events, a floundering response to the need to grow and sustain the destination and acquire tourist arrivals and receipts occurs. Small island destinations can ill afford to flounder in this endeavor.

However, demand techniques can be exercised to assist in understanding the relationship between supply and demand for tourism products. Understanding the demand system widens prospects for measuring elasticity. The concept of elasticity allows for the means to grasp the foundation of the variables being analyzed. This understanding is accomplished through approximations and examinations of the ratio of percentage changes or standard deviations. Demand techniques proffer the opportunity to measure elasticities. Elasticity provides an easily understandable basis of the variables examined by estimating the ratio of percentage changes or standard deviations.[23] A powerful objective is gained in understanding those variables that bear impacts on demand.

That is, there is the possibility of gaining future industry insights that could serve as conduits to promote strategies that better meet demands. Decisions developed from the estimations of elasticity would be beneficial in determining pricing and would encourage the ability to detect tourism revenue trajectory. Considering that increasing prices when demand is price elastic decreases the total tourism revenue, and considering that increasing prices when demand is inelastic expands revenue, a formidable challenge is presented. Nonetheless, the ability to strategize pricing can serve to the destination's advantage as long as potential customers are not dissuaded from their visit to the destination. To maximize revenues via strategic pricing is key to consistency and dominion over a destination's market welfare.

One such technique is the examination of price elasticity. Examining price elasticity for tourism products can provide destinations with an understanding of how price adjustments may impact tourism product supply and demand, and hence revenues earned. Determining the elasticity

---

[23] See, for example, Song, Witt, and Li (2009); Croes (2010).

of demand helps to understand how total tourism revenues may shift based on the offset of a price increase or decrease. For example, when tourism demand is considered to be price elastic, it means that any increase in price will result in a decrease in quantity sold, thus resulting in a decrease in total tourism revenues. On the other hand, when tourism demand is considered inelastic, it means that the market is more tolerant of price adjustments. An increase in price may have little impact on the quantity sold or may reduce the quantity sold. However, the revenue earned by the price increase offsets the decrease in the quantity sold. Thus, inelastic tourism demand will lead to an increase in total tourism revenues.

4.6 TOURISM DEMAND RESEARCH

Despite the rapid growth of international tourism and its importance in many countries, little quantitative analysis has been conducted about small, warm-water destinations. These studies were conducted mostly in developed countries. They suggest that countries can benefit from rising real income. The effect of changes in relative prices and exchange rates varies between countries, indicating that some countries are more sensitive to deteriorating price competitiveness than others. A search from 1990 to 2017 has indicated that tourism demand is an under-researched area in small island destinations, such as the Caribbean region. The scholarly work body is small: with the Scopus database recording only 27 articles related to demand analysis published between Jan. 1990–Dec. 2017. That is one article per year regarding this crucial subject for small island destinations (see Table 4.2).

This dearth of research confirms a similar finding of a study by Shareef, Hoti, and McAleer published in 2008, which found little research into tourism demand regarding small island destinations.[24] The overwhelming majority of studies that have applied quantitative techniques to tourism, for example, in the Caribbean, has been limited and of questionable quality. However, two methods have exhibited a great degree of statistical rigor, namely the time series (non-causal) approach and the econometric (causal) approach. The causal approach better suits the requirements of enhancing the understanding of the predicaments of determining the flow of tourists

---

[24] We already noted in chapter one that small islands are an under-researched subject.

and identifying tools to guide marketing and competitive alternatives to maximize revenue in times of both excess supply and excess demand.

TABLE 4.2 Selective Journal Publications about Tourism Demand in the Caribbean Area from 1990–2017

Authors	Journal
Croes and Ridderstaat (2017)	Tourism Economics
Ridderstaat and Croes (2017)	Journal of Travel Research
Ridderstaat and Croes (2015)	Journal of Travel Research
Ridderstaat, Oduber, Croes, Nijkamp, and Martens (2014)	Tourism Management
Mwase (2013)	Applied Economics Letters
Onafowora and Owoye (2012)	Tourism Economics
Korstanje and Clayton (2012)	Worldwide Hospitality and Tourism Themes
Buckley (2011)	AMBIO: A Journal of the Human Environment
Bresson and Logossah (2011)	Tourism Economics
Pentelow and Scott (2011)	Journal of Air Transport Management
Schubert, Brida, and Risso (2011)	Tourism Management
Henthorne and George (2010)	International Journal of Business Insights and Transformation
Jackman and Greenidge (2010)	Tourism and Hospitality Research
Moore (2010)	Current Issues in Tourism
Besco and Scott (2010)	Tourism and Hospitality Planning and Development
Vanegas (2009)	International Journal of Tourism Research
Padilla and McElroy (2007)	Annals of Tourism Research
Croes and Vanegas (2005)	Tourism Management
Maloney and Rojas (2005)	Applied Economics Letters
Shareef and McAleer (2005)	International Journal of Tourism Research
Huybers and Bennett (2003)	International Journal of Tourism Research
Greenidge (2001)	Annals of Tourism Research
Levantis and Gani (2000)	International Journal of Social Economics
Vanegas and Croes (2000)	Annals of Tourism Research
Dalrynple (1999)	Annals of Tourism Research
Yoon and Shafer (1996)	Journal of Travel Research
Carey (1991)	Atlantic Economic Journal

The consistent maximization of revenue opportunities is required to sustain the advantage of a destination's tourism industry. Thus, pricing is a crucial component to that end. However, pricing is a worrisome task. At its zenith, it becomes strategically favorable for drawing tourist arrivals; but it can also dissuade would-be arrivals given their dismay over considerably disappointing prices.[25] As destinations grasp the impact and value of calculated pricing strategies and their integrated relationship with sustained advantage, they can premeditate their prospects to maximize potential revenues. For example, if both methods forecast that hotel room sales will fall drastically next year, a presumably alarmed manager of the hotel can cut prices and increase the advertising budget to offset the fall on sales. If he now asks both forecasters how these changes will affect their forecast, the time series analyst can only answer that his forecast will remain intact because those factors have not appeared in his equation. The econometric approach attempts to explain tourism demand concerning one or more variables to forecast future demand. In a final analysis, however, it is the quality of the decision's outcome that ultimately matters the most.

If market information is inadequate or incorrect, then policies regarding demand must also be inadequate or incorrect. Thus, information regarding market demand must be founded on empirical evidence to ensure that decision-making bodies have the greatest opportunity to make policy based on reliable and valid data. This is vital as the tourist industry's economic welfare and the forward growth of the destination are dependent upon accurate and edifying data. Such data allows tourist businesses to develop in a manner conducive to attracting spending tourists to the destination. Spending tourists contribute to the overall welfare of the destination, including local businesses, local employment, allotments earmarked for education, health, and infrastructure development. In effect, the destination experiences the overall benefits of development based on conclusive information afforded via scientific evidence. Benefits for the destination are evident under this circumstance. Islands, such as those of the Caribbean Basin facing poverty due to their growing population needs, could border on a road to catastrophe or a road to prosperity depending on the worth of the information they receive.

Estimates of future demand may indicate how the Caribbean countries can cooperate in the promotion of regional tourism: and furnish

---

[25] See, for example, Song, and Wong (2003). For a more recent account of tourism demand, see, for example, Dogru, Sirakaya-Turk, and Crouch (2017); Croes and Ridderstaat (2017).

empirical knowledge for informed decisions to further regional economic and tourism integration. However, this cooperative manner is premised on the same conventional strategy to pool financial resources to market the region. Instead, a new approach should be centered on leveraging greater empirical knowledge and understanding of consumer trends to make local businesses more efficient and innovative. Thus, qualitative information and analyses of data could more effectively be intended to overcome small island constraints.

4.7 LESSONS FROM TOURISM DEMAND ANALYSIS

The few studies that have been done utilize the econometrics approach to explain tourism demand, such as Barbados and Aruba.[26] These studies point out that income in the tourist originating countries explains the changes in arrivals, with varying degrees depending on destination. For example, Aruba experienced income elasticity of demand of 1.43, 2.52, and 1.82 from the United States (USA), the Netherlands, and Venezuela, respectively (Croes and Vanegas, 2008). Greenidge (2001) reports elasticities from Barbados, yielding 1.512%, 2.268%, and 3.134% from the United Kingdom (U.K.), USA, and Canadian markets. However, caution is required because income elasticities may not remain stable over time due to uncertainty induced by economic fluctuations.[27]

Three standards have traditionally served to illustrate tourism demand-pricing, time, and income. However, how these standards have influenced the industry has undergone a reconsideration of their manipulative power. This reconsideration could bear a marked effect on small island destinations in terms of their operations to build and sustain their tourism industry. Indeed, studies have submitted that income elasticity could be compounded by circumstances that cause its decrease. Then, the point here is that where tourism once boasted a characteristic portrayal of effortless extravagance and opulent accessibility, its allure was diminished.

---

[26] See, for example, Clarke (1978); Metzgen-Quemarez (1989); Carey (1991); Croes, (2010). Clarke (1978) pointed out 20 years ago is still true today: studies regarding the demand for international tourism are still lack, particularly studies on small countries.

[27] An increase in destination choices, money illusion, and human psychology may trigger income elasticity instability. Money illusion and human psychology influence how income is consumed, thereby spawning business cycles affecting capacity utilization and altering tourism quality of service, opportunities for market share, industry profits, employment, and the overall welfare of the host country and its residents.

The instability of demand elasticities progresses via the income of the tourist's origin country, exchange rates, and the tourism product's pricing-all of which are elements of businesses that are cyclical by nature. Tourism demand studies suggest that elasticity is not static but assumes a dynamic movement in keeping with various changes in business cycles. Such studies suggest that the relationship between business and tourism demand cycles may be disproportionate and may include volatility in income and price elasticities. By way of example and according to Moore and White-hall, tourism demand for Barbados had maintained a distortion between the years 1957–2002.

Modifications, transformations, or variations in economics, whether growth-enhanced or reduced, could impact the source country's demand and tourist demand for that which tourism could proffer.[28] As a result, where once the tourist arrivals may have boosted long-term upward trends, uncharacteristic deviance may swing downward. Such a change could infringe on tourist product preferences and destination decisions. Moreover, the uncharacteristic deviance also referred to as series cycles, could occur in manners that are unclear or that require time to unfold. According to Smeral, tourists' social status could foster a delayed reaction to the cycles. That status is simultaneously cultivated via tourists' patent consumption.[29]

Demand elasticity can fluctuate insofar as income via wages, benefits, investments, everyday expenditures, and job security can also fluctuate. Some mechanisms contribute to decreases in demand elasticity, such as the greater availability of destinations from which to choose. Besides, the manners in which prospective consumers conceive their disposable income and their actual utilization give rise to an impact business cycles. The trajectory of business cycles, then, bears a distinct impact on the tourism industry in terms of the bottom-line profits, the manners in which the locals do or do not experience the advantages that the tourism business could offer, a reduction in tourist offerings, less than optimal service to consumers, and ultimately static progress. In terms of the Caribbean, the shift in business cycles of more established and progressive economies much further in their economic development could impact tourists' ability to visit a destination. This

---

[28] See, for example, Guizzardi and Mazzocchi (2010); Mayers and Jackman (2011); Smeral (2012).
[29] See Smeral (2012).

impact could then filter to the Caribbean and gravely influence that region's economy.[30]

While growth periods rise and fall, tourists adjust their travel intentions. The tourist may delay travel to distant locations or cancel travel plans altogether. Conversely, as growth rebounds, the tourist may resume plans to travel.

The evidence about the consequences of changes in price is inconclusive, according to the tourism demand literature. Some of the studies suggest high price elasticity, such as Rosenzweig's study of 1988. This study found high intra-Caribbean elasticity of substitutions for tourists from the U.S. of 1.33 and 2.45 for tourists from around the world. Alternatively, in a study I did in 2010 found low price elasticity, more in line with studies of the United States, Canada, Mexico, Italy, France, Spain, Turkey, and South Korea. These seem more plausible than Rosenzweig's findings because the price level may be the outcome of product differentiation or uniqueness and price strategy. The former is reflected in the 'homey' touch to the visitors' experience, the destination's activities, and the type of interaction between host and guest.[31]

Prices usually are not readily subject to change, and, because of the global condition, tourism prices are related to the external markets. Any eventual increase in the total cost component of a destination will most probably be in the profit level rather than the price level. However, what may lie at the bottom of Caribbean tourism effectiveness and competitiveness is the price elasticity of demand. Basing decisions on the elasticity estimation can be a beneficial exercise in the set-up of a pricing strategy. Because changing prices is not an easy exercise, setting the optimal price has become a strategic advantage of any destination in the marketplace, particularly if it can avoid discouraging potential customers. By creating this strategic capability, a destination avoids being consistently behind the curve on pricing and missing meaningful opportunities to maximize revenues in times of both excess supply and excess demand.

---

[30] For a thorough discussion on the influence of business cycles on demand elasticity, see, for example, Croes, and Ridderstaat (2018).

[31] Croes (2010) claimed that more than 50% of the tourists coming to Aruba have visited Aruba more than twice and that 90% of these are willing to recommend Aruba to friends, relatives, and acquaintances. This high loyalty ratio can significantly impact the total economic output of tourism in Aruba, as suggested by Reichheld and Sasser (1990). They empirically demonstrated a positive relationship between customer loyalty and profitability.

These missed opportunities may explain concerns expressed by the studies of Ceata-Hatton in 1997 and Bryan in 2001 related to the 'negative externalities' present in the industry today, such as the devaluation of the destination, unemployment, and underemployment, a growing presence of prostitution and drugs, increased violence and ecological imbalances. The policymaker faced with these pressing problems may be tempted to leap to the 'easy' answer to increasing tourism to increase foreign exchange and decrease unemployment. Suppose he fails to do so in a controlled manner. In that case, however, distortions in growth and result in damage to the ecology could occur, as well as excess carrying capacity, and overuse of public utilities. Indeed, as small destinations have strived to maintain and grow their market share of tourists, they have sought to fulfill the social amenities and conveniences that tourists preferred, making their destination competitive with the other Caribbean and global destinations.

However, in so doing, they neglected to recognize the toll to their environment and resources that their strategy to meet such preferences extracted. The result was fertile ground for locals' resentment. Further, small island destinations were persistent in using their factors of production to augment their tourism industry. The effect was to create a path to diminished tourist arrivals once the incentive for them to visit paradise was de-energized by the very development that sought to capture them in the first place. Thus, this symptomatic weakening of the Caribbean's allure has had effects of reduced competitiveness.

Without robust, empirical evidence focused on examining existing and future tourism potentialities for the Caribbean region, small island destinations risk the types of negative externalities that render the destination unattractive to tourists and damaging to the cultural and social welfare of the locals. Negative externalities that include criminal activity, destructive impacts to the natural resources, and economic issues diminish the potential of the destination. As tourism prospects fade, local employment issues arise. Besides, some studies have indicated that external shocks such as Dutch disease could also contribute to a declining, small island tourism industry and the possible high elasticities that could occur to compensate for lost revenues. Thus, an empirical study's advantages in forecasting variables could influence the small island tourism industry's welfare to become apparent. Indeed, without empirical data to aid strategic planning, small islands risk submission to doubts in their ability to grow and sustain themselves.

Only with increased and improved information on tourism demand, its market sources, and the spending characteristics associated with it, can the governments in the Caribbean take maximum advantage of the tourist export component in their national plans for economic growth and development. Typically, arguments for allocating more significant resources to the tourist sector to increase income levels have not been based on empirical analyzes. The existence of a reliable and well-researched plan demonstrates that the government, at a minimum, has analyzed the market situation and has designed a program of investment in the tourism sector based on its forecasts of future demand. Besides its due diligence, a potential investor has the additional assurance that the host government is a knowledgeable partner in terms of approaching the tourism sector in a sophisticated and businesslike manner. There is a need to measure the effectiveness and advertising efforts beyond increased market share. Tourism marketing organizations are increasingly asked to account for programs' effectiveness via objective measurements that could reveal specific economic and social impacts such as job creation, tax revenues, and investments.

4.8   THE TOURISM DEMAND ANALYSIS PUSH

Unquestionably, tourism research has grown in importance across the Caribbean and the world over the past two decades, indicating that the tourism businesses that prosper are those that use market research to chart their course of action. The Caribbean countries face the necessity for drastic improvement of their entire tourism operations if they are to compete successfully in the global tourism marketplace. Despite the rapid growth of international tourism and its importance in many countries, little quantitative analysis has been conducted. These studies were conducted in small island destinations. Only a handful of studies exists that quantitatively assesses tourism as a benefit to small island destinations, such as the one recently published by Nunkoo and other co-authors about tourism and economic growth.[32]

As market research reveals that tourist demand encompasses experiences distinctive to tourists' desire that such experiences be inimitable encounters, destinations find they must develop the types of offerings that

---

[32] See Nunkoo et al. (2020).

satisfy the wants and needs of the tourist-following the tourists' purposes for traveling to the destination. If the destination's offerings support the tourists' concept of what constitutes a unique and satisfying experience, then the destination comes to recognize that the tourists will pay for the opportunity to immerse themselves in the destination's goods and services. Thus, small island destinations must be compelled to accept that continued research into tourism demand is essential to growing and sustaining its market.

Although studies regarding the characteristics and elements of tourism demand have provided information on the factors that define tourism flow, they have disregarded the need for necessary and adequate empirical evidence on TS's sustainability. Whether or not the tourism industry's development is sustainable, or whether its developmental process changes over time, or how that change impacts tourism over time, or how demand elasticities ebb and flow over time require empirical data. Such temporal concerns regarding tourism in small island destinations may be provoked by the exhaustive exploitation of its main marketable feature, i.e., its natural resources.[33] At the point at which this one feature has reached its ceiling, growth becomes dependent on the productivity of labor. This shift may result in slowing tourism growth. Thus, given this view, it may not be possible for tourism, even at its zenith, to be a long-term, competent pathway for development.

Small islands realizing TS as their pathway to prosperity do not have operative control over global tourism demand according to how that demand is shaped by the prolific and expanding wealth of source countries. However, they may respond effectively and exert control over tourism demand by gearing their marketing efforts towards intelligent pricing strategies and the right market segment. To enable the tourism industry to compete effectively, a better understanding of the demand elasticity is required.[34] Importantly, tourism demand is concerned with the essential factors in determining and predicting foreign demand for small islands.

Understanding the expected market demand is fundamental and should have the highest priority. Most tourism development programs fail to do so,

---

[33] For an excellent discussion if tourism can propel sustainable development, see Candela and Cellini (2006); and Marsiglio (2017).

[34] By better managing pricing policy to maximize revenues, particularly in time of slump (seasonality), it may also prove beneficial for some states in the United States. The economy of Florida, for example, is very much intertwined with that of the Caribbean islands. Florida is a major supplier of goods and services to the region and serves as a central hub for flights to the islands.

however, because they have not adequately evaluated the relative importance of income, relative prices, exchange rates, and the impact of business cycles; or they have not regularly assessed the one non-economic event of great interest, such as the reintegration of Cuba into the North American economic order. The international tourism demand for small islands depends upon changes in these variables. Without representative and accurate international demand estimates, both public and private sectors cannot invest scarce resources efficiently. In this context, the research studies' findings will provide essential insights for small islands' policymakers in formulating policy and for the entrepreneurs in running their businesses. Decisions to be made on prices, promotional or strategic marketing programs, distribution, and allocations of human, natural, and capital resources all require reliable predictions of current and future demand trends. Researching the existing and potential markets will facilitate a much more intelligent approach to supplying those markets' needs and wants.

I will examine the determinants that shape the relationship between TS and economic growth in the next three chapters.

KEYWORDS

- **economic growth**
- **embrace tourism**
- **gross domestic product**
- **tourism demand analysis**
- **tourism effect**
- **World War II**

CHAPTER 5

TOURISM SPECIALIZATION AND ECONOMIC GROWTH

In the previous chapter, I anchored tourism development on a demand orientation because a tourism demand orientation conceptually matters for economic growth. More demand through higher tourist arrivals is associated with economic growth in two ways. First, tourists trigger spending at a destination. They spend on lodging, meals, attractions, shopping, transportation, souvenirs, and entertainment. All this spending goes directly to firms (investments and profits), households (through wages, tips, or services), and government. The latter receives a portion of the spending through fees, taxes, and goods and services. However, spending also affects other sectors in the economy. That is, tourist spending goes beyond hotels, restaurants, transportation, and amenities. Tourists bring with them money spent by local firms on traded sector production, such as food production, drinking water production, clothing, and printed media.

Second, tourist arrivals through demand may have an indirect effect on the GDP through productivity spill-overs. For example, productivity spill-overs occur when international hospitality firms establish themselves in a destination bringing high skilled staff and knowledge to the destination. The influx of knowledge and experience contributes to the destination's tourism growth. Further, the firms could sponsor and interject new ideas and the types of knowledge that could be used to expand the destination's economy. These new ideas could be related to management processes to improve productivity, a new product or service, or a new market segment.

The demonstration effects of firms can influence productivity levels. For example, new firms may use technology diffusion and entrepreneurial learning to improve the quality of their products and services. There succeeds from these demonstration effects a web or network of influence over local entrepreneurs to meet the same type of standards and quality of their goods and services to which large international firms and chains must aspire and maintain themselves. The impact becomes one of local, national, and international weight as related tourist service entities raise their abilities and qualities to meet the tourist industry's required demands, doing so in a standards-based manner.

Remarkably, mainstream economics has not considered tourism as a determinant of growth. Rather, tourism has been a low productivity economic activity which by itself could not propel economic growth. Yet, investment, and job opportunities for any tourist destination rely on growth potential. Therefore, understanding the determinants of growth is crucial in any discussion of tourism as a viable economic structure to achieve prosperity. This chapter will discuss the theoretical foundation of justifying tourism as a growth vehicle and will validate it through empirical analysis.

5.1   KALDOR AND ECONOMIC SPECIALIZATION

In the previous chapter, I contended that the consumer theory of choice explains tourism demand. This theory is based on consumers' preferences and how these preferences determine destination choice. However, this theory does not explain why countries specialize in a particular economic activity such as tourism. Therefore, the theory lacks the supply component of the tourism activity. The relationship between demand captured through tourism arrivals and economic growth is conceptually undergirded by Kaldor's writings (1966, 1970, 1985). Kaldor asserts that growth results from a demand induced process. By his account, growth stems from the expansion of manufacturing (characterized as the engine of growth). This expansion is the outcome of the interaction of the dynamics of efficiency based on learning by doing, technology, and demand.

Kaldor's view stressed the need to focus on demand and the sector with the most promising growth opportunities. His demand focus prompts examining promoters, facilitators, or inhibitors of demand to assess growth. He was one of the first to focus on the relevance of sectoral development. Kaldor eschewed the resource-constrained approach enshrined in the supply-side studies. He provides the conceptual foundation to focus on tourism demand.

The Kaldorian approach asserts that foreign demand is the promoter of economic growth, and the growth rate of exports determines the rate of output of growth. Kaldor recognized the relevance of exports for extending the market. He also posits that export per se does not conjure growth. Instead, economic growth relies on the sectoral approach. Kaldor's approach suggests that what a country exports matters; that is why it is crucial to inquire into the structure of production and exports. That what a country exports matters seems to deviate from the classical Ricardo's comparative advantage theory.

As mentioned in Chapter 2, Ricardo's theory provides the foundation for countries to trade. His justification for trade is anchored in the relative productivity advantage, suggesting that differences in national productivity spawn gains from trade. The implication is that every country enjoys a comparative advantage just by being different. It does not matter what a country specializes in; a country can benefit equally, either by producing apples or Apple computers. Ricardo's policy implication was managing opportunity cost: export goods and services that a country can produce at low opportunity cost while importing those that a country can produce at high opportunity cost.

Kaldor differs in this respect from Ricardo. What matters is what a country export. Moreover, what a country exports are related to the economic sector's growth characteristics. Not all economic sectors are the same: some have more growth potential than others, and the defining attribute is if the sector grafts increasing or diminishing returns. This is an essential conceptual recognition because this perspective provides a pathway to how small islands can grow their domestic market. If tourism can propel increasing returns, then tourism could spawn economic growth on a small island despite size constraints.

Kaldor's approach provides a clear pathway for small islands if they want to prosper. This pathway involves four main components. First, small islands should practice a development strategy premised on a demand orientation. Tourism fits this first norm. Tourism can spawn a significant inflow of foreign exchange that can pay for imports to further industrialization. For example, the study of Balaguer and Cantavella-Jorda published in 2002 revealed how openness to the outside world and an export orientation through international tourism in the 60s helped Spain's rapid industrialization by paying for consumption and investments. Tourism with Spain was picked as the leading sector to boost economic growth because tourism as a sector had the most favorable growth characteristics. So, export growth through international tourism demand has both demand- and supply-side effects conducive to growth.

Second, trade, and export can increase demand. As tourists frequent a destination, they stimulate and manipulate demand for goods and services. This, then, pushes the thresholds of destination production in an outward trajectory. Thus, as demand gives rise to production, production responds to increased market availability, extending, and expanding its goods and services' original borders. The broader market allows the destination to take advantage of economies of scale and opportunities for the product balance afforded via specialization and diversification. The tourism sector can propagate upstream linkages when, for example, a hotel buys inputs from suppliers to offer its services and goods to the tourists. For example, these suppliers, such as advertising and bakeries, buy their inputs from other suppliers to create their goods or package their services to satisfy hotel demands. Hotels may also sell their products to non-tourists, for example, local businesses, which may spawn strong downstream sales as they become additionally used as intermediate inputs in further production.

Similarly, other tourism-related industries, such as bakeries, grain product manufacturing, and support activities for transportation, may also spawn forward linkages when goods are produced for tourism use. Tourism also can provoke diversification because tourism facilitates learning and is a channel for local product promotion. For example, Lejarrana and Walkenhorst's study in 2007 indicates that local entrepreneurs connect with and become more participatory in tourism type activities due to the tourism industry. In small island destinations, the tourist sector allows local businesses to absorb and utilize cost-effective means to meet tourist demand. Moreover, as local tourist sectors affiliate with others in the local economy, the resultant commerce gives rise to ongoing economic expansion.

Arguably, the tourists' presence reveals a unique situation where the local entrepreneur, with little effort and cost, can engage in a discovery mission to discern tourists' tastes and preferences and determine the corresponding product to satiate their tastes. This entrepreneurial process of cost discovery enables the opportunity to use tourism as a potential economic diversification platform. We encountered this in the Aruba case, after its 20 years of tourism development, when an interesting burgeoning of cross-pollination was displayed. A report by the Aruba Central Bureau of Statistics looked at the employed population's economic activity distribution from 2000 to 2007. The report found an increase in the employed population gravitating towards agriculture, hunting, and forestry, manufacturing, construction, real estate, and health. These sectors grew respectively by 97%, 83%, 67%, and 60%.[1] It seems that tourism specialization (TS) could spark a U-curve pattern in its relationship with GDP. As its GDP per capita rises, the structure of the production of goods diversifies launching new products and through diversification within those goods that are already being produced or exported. The Balearic Islands reveal a similar diversification pattern. These islands also show a diversification trend away from its TS pattern decreasing its specialization index for against the national structure during 1986 to 2002.[2] These examples underscore why developing countries engage with tourism as a diversification strategy away from tradable goods that lost their competitiveness in the global market.[3]

Third, specializing in the economic sector reveals increasing returns. Tourism expands the economy through increasing ToT. For example, the study of Lanza and Pigliaru of 2000 demonstrates that specializing in an economic sector with lower productivity gains does not necessarily lead to sluggish economic growth. The condition to propel economic growth through tourism is when the substitution elasticity between tourism services and manufacturing goods is less than one; tourism is highly valued over cheaper manufacturing goods. Therefore, economic growth may be steadily driven when tourism goods and services are continuously appreciated. Notably, tourism's characterization as an instrument for sustained growth is feasible, providing its ToT constant and upward movement. That feasibility also requires that tourism revenues be consigned to the purpose of acquiring imports of capital goods. This condition is referred by Brau et al. in their study of 2007 as the 'optimistic interpretation' explaining why tourism is a growth vehicle.

Further, and fourth, increasing returns activities are associated with high-income elasticity of demand in source markets. Smeral asserts that the production costs of tourism will continue to grow over time, inducing higher prices.[4] When tourists visit a destination, they consume local amenities and non-tradable goods such as scenery, weather, heritage, and culture, nightlife, restaurants, hotels, attractions, and shopping. These higher prices can last because converting these non-tradeable goods into tradeable can bestow monopoly power to the destination. Insofar as no destination is

---

[1] See CBS (2008). Development of Aruba's Labor Force 2007. Oranjestad: CBS.
[2] See Buswell (2011).
[3] See, for example, Lejarrana and Walkenhorst (2007).
[4] See Smeral (2003).

identical or similar, each is distinguished due to its singular amenities, attractions, and offerings. This lends opportunity for the development of cost advantage.

Gaining cost advantage appears difficult in terms of tourism production. However, as more appealing products can gain cost advantage, it is likely that gaining cost advantage rests with the demand side of the tourism product-the greater the appeal, the greater the demand. The more appealing the product is, the greater the ability to assign and acquire higher prices and to continue to reap those prices over time. However, a specialization-growth perspective may be corrosive if it relies on demand-side factors that relish in price elasticity. Competing on price is like a race to the bottom in a supply-constrained environment, such as a small island. Therefore, a tourism product that stems from higher-income elasticities has a greater impact on economic growth. The requirement is that tourism remains a luxury good to sustain growth over time.[5] The higher the income elasticity and the lower the price elasticity, the more substantial the price increase to beat tourism production costs. However, because tourism demand is dynamic, depending on the changing nature of tourists' tastes and preferences, the income elasticity may change. Keeping an eye on the income elasticity level and magnitude is a critical destination management requirement.

5.2   THE TOURISM-LED GROWTH HYPOTHESIS (TLGH)

The search for a causal relationship between tourism and economic growth is a recent phenomenon. In 2002, Balaguer and Cantavella-Jorda published an article showing a positive causal relationship between tourism and economic growth, running from the former to the latter. This study postulated the TLG hypothesis in the tourism literature. The tourism-led growth (TLG) anchors two main theoretical foundations: The Keynesian multiplier and the Lucas' endogenous growth model. The former posits that tourism is an exogenous component of aggregate demand conveying a static perspective of tourism development. Based on the elasticity of substitution between tourism and manufacturing, the latter claims that tourism is dynamic and growth-enhancing in the long run as long as the substitution elasticity is smaller than one.[6]

---

[5] See, for example, Keane (1997); Croes (2011); Alvarez-Albelo and Hernandez-Martin (2009).
[6] See, for example, Copeland (1991); Hazari and Sgro (1995); Lanza and Pigliaru (1995).

Three propositions follow from the TLG theoretical foundations. The first proposition claims that there is a positive relationship between TS and economic growth. Moreover, second, the direction of the relationship is from TS to economic growth. And third, small islands have a comparative tourism advantage, and trade enlarges the small island domestic market through increased demand for international tourists.[7] The growth potential of tourism is revealed through increased terms of trade (ToT). The empirical evidence corroborates these theoretical propositions because small islands, despite the endogenous growth theory's pessimistic predictions, grew.[8] The combination of two elasticity phenomena (high-income elasticity and price elasticity ambiguity) contributes to relatively stable export earnings of tourism products compared to commodity groups benefiting the ToT of destinations specializing in tourism.

In 2003, Manuel Vanegas and I published a study in which we validated the TLG in Aruba's case.[9] Other studies, including Durbarry, assessed the link between tourism and economic growth in Mauritius's case and found that tourism and economic growth are positively related to tourism triggering economic growth.[10] However, before 2010 only a handful of studies focused on validating the TLG hypothesis in small island destinations.

Brida and Pulina, in their study of 2010, assert that the literature has neglected tourism because tourism is a different type of export than agriculture and manufacturing-meaning that because the tourist must visit a destination, the visit constitutes an export. Tourism production and consumption take place simultaneously in the same location. About tourism, the factors of production do not dictate the creation of saleable goods. Rather, the economic benefits of tourism are shaped and generated by the consumer. Indeed, as tourists disclose their preferences for goods, services, and experiences, they exude a dynamic influence on the production process itself-not only for tradable goods but for the non-tradable (historical structures, Sun, Sand, and Sea) that may primarily

---

[7] See, for example, Algieri, B., (2006). International tourism specialization of small countries. *International Journal of Tourism Research*, 8(1), 1–12.

[8] See, for example, Lanza, A., & Pigliaru, F., (2000). Why are tourism countries small and fast-growing? In: Fossati, A., & Panella, G., (eds.), *Tourism and Sustainable Economic Development* (pp. 57–69). Dordrecht: The Netherlands: Kluwer Academic Publishers; and Brau, R., Lanza, A., & Pigliaru, F., (2007). How fast are small tourism countries growing? Evidence from the data for 1980–2003. *Tourism Economics*, 13(4), 603–613.

[9] See Croes, R., & Vanegas, M., (2003). Growth, development, and tourism in a small economy: Evidence from Aruba. *International Journal of Tourism Research*, 5(5), 315–330.

[10] See Durbarry (2004).

have drawn them to the destination. The price-setting for both types of goods is derived by tourists' demand, which recognizes tourists as a central construct in the production process. Diverse ToT emerge from tourist demand as that demand builds unique trade milieus for local products.

The foundation and support for the positive relationship between tourism and its contribution to economic growth on a long-term basis are reasoned within the TLG hypothesis. The TLG hypothesis stems from the export-led growth hypothesis postulated by Kaldor, which contends that expanding export can also be a driver of economic growth next to increasing labor and capital within an economy. Exports affect economic growth through efficiency improvements related to competition. For example, my study of 2006 reveals how an increase in Aruba's international tourism arrivals has generated more competition through the presence of new entrants. New entrants more than doubled from 1,940 in 1987 to 4,442 in 1999, a 229% increase, according to Aruba's Central Bureau of Statistics.

Exports also increase the levels of investment that bear an important contribution to demand and economic growth. Tourism promotes investments in new infrastructure, human capital, and competition. Tourism flows to a destination and requires services, attractions, facilities, and infrastructure. These supply units, which are essential amenities fostering demand at a destination, rely on public and private investment. The quantity and quality of these supply units depend on the investment level to create, support, and nurture the destination appeal and image.

Further, human capital is crucial for delivering quality offerings and services, which create a memorable experience. Creating a memorable experience depends on a talented workforce that relates to skills, education, and professional training. As demand increases, the spectrum of job opportunities increases, converting tourism activities as an essential source of new jobs. For example, tourism generates one in seven jobs in the Caribbean. The presence of tourists also means that foreign exchange is generated, which may be used to pay for imports and maintain a minimum international reserve level.

Tourism is vital for small islands because it triggers economic growth. But what is even more critical is the fact that tourism provokes fast economic growth. Empirical evidence reveals a positive relationship between tourism and economic growth, according to several studies. These studies examined this relationship in the context of small islands and indicate, for example, that the top 10 nations ranked according to tourism contribution to GDP are all small islands.[11] Small islands ranked high when considering their tourism performance, but the empirical literature also asserts they grow fast when specialized in tourism. Examples are Aruba, St. Lucia, Cayman Islands, and the Bahamas.[12] Other studies, including Hazari and Sgro's study, also showed TS as a pathway towards economic growth.

They built a dynamic model and claimed that tourism spending would positively impact a small economy's long run growth. They posited that tourism demand altered the locals' consumption pattern by consuming now rather than later via lower saving rates requirements. Also, the study of Lanza and Pigliaru came to a similar conclusion: tourist specialization of a small country positively affects its economic growth. They indicate that tourism appears as growth-enhancing despite the size of the country. Another study that examined 19 island economies concluded that tourism triggers economic growth and tourism development has comparatively higher growth effects than developed and developing countries.[13]

Given that tourism facilitates growth and materializes as a potential economic premium, available resources to fund and support the tourism industry should cascade in that direction more so than other financial areas, as suggested by the TLG hypothesis. Consider an island of 100,000 inhabitants with its economic constraints imposed by the size of this market. Now consider the same island with 100,000 with an additional transient population of 1 million with sufficient discretionary income. The compelling economic growth opportunities through jobs, business opportunities, and enhanced quality of life become more meaningful and possible. Export growth triggers demand in other economic sectors because it can pay for the import content of consumption, investment, government expenditures, and exports. Thus, the TLG hypothesis is a powerful economic opportunity for small islands because small islands have a comparative advantage when specializing in tourism. Let us consider the Caribbean as an illustration.

---

[11] See, for example, Schubert et al. (2010).
[12] See, for example, Brau et al. (2007); Sequeira and Nunes (2008); Vanegas and Croes (2003). Brau et al. (2003) compared the growth performance of 143 countries concluding that countries more specialized in tourism grow faster than non-specialist countries.
[13] See Seetanah (2011).

5.3   ASSESSING COMPARATIVE ADVANTAGE IN THE CARIBBEAN

The Caribbean traditionally has been a geographical area comprising many small countries highly reliant on international trade and heavily dependent on a few exports and export markets. These countries suffer growth costs associated with their domestic market's small size, which renders them particularly vulnerable to numerous economic risks. For example, business cycles in source countries have spill-over effects on island destinations exacerbating the volatility level of tourism demand flows with damaging service quality, market opportunities, profits, jobs, and welfare. From economic growth to economic slowdowns, changes may affect aggregate demand in the source country and potentially affect tourism offerings.

Tourism demand may respond to business cycles with some delay, either constant or showing variation. The interaction between business cycles and tourism demand cycles may exhibit unstable income and price elasticities.[14] For example, a study by Moore and Whitehall published in 2005 found that Barbados' tourism demand responses have been asymmetric from 1957–2002. However, Ridderstaat and Croes' study of 2015 published potential issues regarding the efficacy of using business cycles to analyze tourism demand. The flaw occurs in that business cycles are characteristically tenuous but contribute to GDP. Thus, consideration of timeliness of GDP data (whether used quarterly or annually) becomes an issue. In brief, as business cycles evolve and devolve according to time, GDP data can deposit undue analytical weight. Also, it must be considered that the service the industry affords continues to grow in successful destinations. As the industry grows, its increasing impact on GDP lessens the impact of manufacturing data-where such industrial production had once occupied a more significant role in its bearing on GDP via business cycle analyzes. These features disallow efficient use of business cycle data that would be used as a strategic tool to forward tourism in the Caribbean region-an important consideration given the growing importance of tourism in the Caribbean.

The tourists' presence in the Caribbean has generated enormous economic windfalls. In a 2006 article, I asserted that tourism has been the

---

[14] See, for example, Gouveia and Rodrigues (2005); Guizzardi and Mazzocchi (2010).

most important foreign exchange earner of 16 of the 30 members of the Caribbean Tourism Organization (CTO) and that the Caribbean is the most tourism-dependent region in the world.[15] By 2009, tourism accounted for 15% of the CTO's members' GDP and taxes, 22% of capital investment and exports, and eight jobs (WTTC, 2009). In 2010, they received 19.5 million overnight international arrivals, which equates to a 2.1% share of global tourism and a 13% share of international tourism in the Americas. This means that 30 CTO member countries with less than 1% of the global population attract 2.1% worldwide tourism. By 2019, the Caribbean region received 26.5 million overnight international arrivals, equal to 1.8% of the global tourism share, and generated US$ 34.6 billion in international receipts, equal to 2.3% of the global share. Tourists spent US$ 23.3 billion in the Caribbean in 2010, which is 2.3% of total international tourism global spending.

The Caribbean region tourism performance in the last decade suggests a slight drop in arrivals' market share, while holding a stable market share in international receipts. This performance indicates that the spending per arrival has increased over the past decade. With the Caribbean being so dependent on tourism, it is critical to assess how well the industry performed compared to other regions in the world. Compared to other regions globally, the performance of Caribbean countries from 1986–2001 was significant, outperforming the other regions with the highest elasticity arrivals. The Caribbean region also experienced stable tourism income surpassing the world average by a factor greater than two in income stability. Indeed, as of 2018, one out of every eight Caribbean jobs has been disclosed related to the tourism industry. Future forecasts place that number to one out of every seven by the year 2025. In 2019, the Caribbean was the best performed region in the Americas, according to the World Tourism Organization (WTO).[16]

Given that this data equates to 2.8 million available jobs gives some revelation about the Caribbean's successful dependency on the tourist industry. Further, the World Travel and Tourism Council (2015) indicates the economic impact for the 2025 Caribbean region as high as US$ 74 billion, with tourism receipts ascending beyond US$ 25 billion. Moreover,

---

[15] See, for example, Croes (2006).
[16] See UNWTO (2020). *International Tourism Highlights.* https://www.e-unwto.org/doi/epdf/10.18111/9789284422456, retrieved June 20, 2021.

such data contributes to the Caribbean's display as revealing the world's most significant dependency on tourism.

We already alluded to the opportunities that tourism could bring to small islands due to specialization. If these small islands specialize in this comparative advantage, they may grow in competitiveness and trade. Several studies, including the most recent one from Algieri, suggests that the comparative advantage may lie in the small opportunity cost embedded in a smaller scale, typical of small island destinations.[17] It seems that this is the reason why tourism's economic prowess and potential have caught the attention of these small countries in the Caribbean. No wonder that tourism has been embraced as a strategic engine of growth for many small countries-as a way to overcome the constraints of a small market and to foster rapid economic growth. However, I also expressed concerns regarding the Caribbean tourism performance in the realm of foreign exchange generation. The region was increasing its foreign exchange at the expense of the volume of arrivals compared to prices, revealing a declining trend in foreign exchange generation compared to the world average. Globalization has not been kind to the Caribbean's tourism industry.

That is, the context and fluidity of world market conditions and circumstances force destinations to react positively and effectively to meet supply and demand changes on a global scale-a situation to which the Caribbean region has not reacted well to. For example, the U.S. Great Recession of 2007–2009 rammed the Caribbean's tourism industry resulting in severe tourist declines in arrivals, service offerings, and economic issues. Adverse logistics, such as airline flights offered by source countries left the Caribbean region with little recourse.[18] External shocks curtail the effects of tourism growth on economic growth, resulting in declining tourism flows, foreign exchange, and employment opportunities. In spite of this, Caribbean destinations proved their resilience, bouncing back quickly from would be devastating setbacks. Tourism development in the Caribbean often revealed a V-pattern, quick steep decline, and even quicker steep recovery. Indeed, the region showed a 10-year sustained tourism growth since the recession end in 2009.[19]

---

[17] See Algieri, Aquino and Succurro (2018).
[18] See, for example, Laframbroise, Mwase, Park, and Zhou (2014).
[19] The COVID pandemic halted this sustained growth. I will discuss the pandemic in the last chapter.

Sustained growth after a devastating recession aside, it is important to note that while tourism is vital overall to the Caribbean region, differences exist as to tourism's relevance to each destination within the region. Hence, comparative advantage and factor endowment theories can be applied to tourism. Tourists may choose a destination because of cultural heritage. For example, the Caribbean has several UNESCO heritage and natural sites, including the Belize Barrier Reef Reserve System (Natural, 1996), the Historic Area of Willemstad, Curacao, Inner City and Harbor (UNESCO, 1997), Morne Trois Pitons National Park (UNESCO, 1997) in Dominica, and Brimstone Hill Fortress National Park (UNESCO, 1999) in St. Kitts and Nevis. Several Caribbean islands participate in the Slave Route Project, launched by UNESCO in 1994 as an essential heritage of the region. Overall, heritage has become an essential asset for economic development. Another motive for visiting a Caribbean Island is a pilgrimage. An important pilgrimage site is the Basilica of Our Lady of Altagracia of Higuey in the Dominican Republic. Also, Mount Zion, a mystical and holy place in Jamaica, is visited by tourists interested in the Rastafarian Culture. SSS may also attract tourists as the Caribbean is well-known for its beautiful beaches and pleasant weather.

Sinclair and Stabler in 1997 emphasized the relevance of comparative advantage in explaining tourism flows.[20] In Chapter 2, I amply dwelled on the conceptual underpinnings of comparative advantage as the source of specialization. The chapter argued that specialization, and *in tandem*, TS, could trigger economic growth in small island destinations. The ontological foundations of specialization lie with Adam Smith's division of labor perspective. Measuring comparative advantage has been notoriously challenging, as a comparative advantage is characteristically not directly observable. The challenge resides in the distinctiveness of history, traditions, and cultural and natural resources that render the measuring of comparative advantage cumbersome.

Bela Balassa introduced the Balassa index in 1965 to address the measurement challenge. The index is anchored on the principle of revealed comparative advantage, which assumes trade flows reveal the underlying patterns of a country's comparative advantage. The index references the export share of a country's particular commodity (e.g., tourism) to all commodity exports of that country. It relates that outcome to a similar

---

[20] Other studies such as Algieri (2006); and Webster et al. (2007) also indicated the importance of comparative advantage as a justification of tourism flows.

calculation for all selected countries generating that product in that period.[21] A Balassa Index exceeding one suggests that the country is specialized in tourism. Simultaneously, a value below one indicates that the country is less specialized in tourism than the average of the sample of countries.

To understand how the Balassa Index works, I estimated a Balassa Index value for a sample of Caribbean countries. Table 5.1 provides the Balassa Index value for this sample. None of the countries in the sample display a strong comparative advantage. Trinidad and Tobago and the Dominican Republic display a value less than one, which means they displayed a comparative disadvantage in tourism in 2008 compared to the other countries. All the remaining Caribbean countries in the sample reveal a medium to high advantage in TS.[22] This result is promising given the extent of investment undertaken in this sector by the private sector and governments. For comparison purposes, I also estimated the Balassa Index for the year 2016 with the same country sample. The results indicate that Barbados and the Cayman Islands reveal a strong comparative advantage, while most of the other destinations dropped in their scores to weak comparative advantage.

The shifting in time pattern is also reminiscent of a study conducted by Jackman, Lorde, Lowe, and Alleyne, published in 2011. Their study, employing the Balassa method covered 18 small island destinations over eight years from 2000 to 2007. They concluded that these islands had increased their comparative advantage over the sample time period. However, the same study concluded that their comparative advantage was weak.[23]

The conclusion that countries' comparative advantage may be weak is insightful. The strength of the revealed comparative advantage may differ

---

[21] The estimation of the index followed by Algieri (2006). See Algieri, B., (2006). International tourism specialization of small countries. *International Journal of Tourism Research*, 8, 1–12.

$$B_{yi} = 100 \times \dfrac{\dfrac{x_{yi}}{\sum_{y=1}^{N} x_{yi}}}{\dfrac{\sum_{i=1}^{M} x_{yi}}{\sum_{y=1}^{N} \sum_{i=1}^{M} x_{yi}}}$$

where; x is the exports; y is the commodity; i is the country; N is the total number of commodities; M is the total number of countries.

[22] We followed the classification of Hinloopen and Marrewijk (2001), as indicated by the 2018 study by Algieri, Aquino, and Succurro (2018). The classification suggests $0>B<1$ as a comparative disadvantage; $1>B<2$ as a weak disadvantage; $2>B<4$ as a medium disadvantage; and $B>4$ as a strong comparative advantage.

[23] See Jackman, Lorde, Lowe, and Alleyne (2011).

among those countries specialized in tourism, suggesting that averages can conceal uneven advantages among small islands.

To capture these potential differences, I compared these countries by normalizing the Balassa Index (see column)[24] by using a formula based on the sum of exports and imports divided by the trade volume. The normalized index takes a minimum of −1 (the destination depends entirely on imports) and a maximum value of +1 (the destination depends entirely on exports). The normalized index tells us that countries with a larger trade surplus suggest a stronger comparative advantage.

Table 5.1 suggests that we can divide the sample into two groups: the stronger group of countries (identified in bold) consists of Antigua and Barbuda, Aruba, The Bahamas, Barbados, and St. Lucia. The weaker group of countries consists of Dominica, Grenada, Jamaica, St. Kitts and Nevis, St. Vincent and the Grenadines, and Turks and Caicos. The estimation of this index for the year 2016 reveals that the sample of countries, in general, fared worse compared to 2008, suggesting a drop in comparative advantage. Six destinations dropped in the comparative advantage ranking, achieving opposing comparative advantage, namely, Cayman Islands, Dominica, Jamaica, St. Vincent and the Grenadines, Trinidad and Tobago, and Turks and Caicos. The other large drop, while still positive, is the case of the Bahamas. Finally, the estimates suggest considerable improvements in Grenada, St. Kitts and Nevis, and St. Lucia.

5.4 TOURISM SPECIALIZATION (TS) FLASHPOINTS

Even if small islands could show improvements in their comparative advantage, this improvement in itself is not a guarantee that small islands can and will grow. Remember that Adam Smith (1976) taught that the market's extent limits economic productivity and growth. Larger markets have an advantage over smaller markets because larger markets enable economies of scale, entrepreneurial incentives, and technological advances that affect each other. This virtuous circle breeds growth, which produces more savings, makes capital more readily available and affordable and undergirds machinery and labor investment. Such investments are justified only

---

24 $NB_{ji} = \left[\dfrac{x_{ji} - m_{ji}}{x_{ji} + m_{ji}} \right]$

where; NB is the normalized balance; j is the commodity; i is the country.

TABLE 5.1 The Balassa Index for a Sample of Caribbean Countries, 2008 and 2016

Destination	Int. Tourism Receipts (billions)		Balassa Index		Normalized Balassa Index	
	2008	2016	2008	2016	2008	2016
Antigua and Barbuda	0.334	0.332	1.64	1.28	0.09	0.02
Aruba	1.237	1.631	3.04	1.37	0.04	0.05
Bahamas	2.144	2.591	1.86	0.85	0.17	0.02
Barbados	1.194	1.038	1.75	3.15	0.24	0.22
Cayman Islands	0.518	0.686	3.38	0.40	–	−0.06
Cuba	2.258	2.907	1.59	1.65	–	0.20
Dominica	0.076	0.132	1.3	1.20	−0.09	−0.11
Dom. Republic	4.166	6.723	0.95	0.42	–	−0.05
Grenada	0.127	0.149	2.02	0.64	−0.16	0.03
Guadeloupe	0.51	–	2.8	–	–	–
Haiti	0.36	0.504	1.01	2.10	–	−0.32
Jamaica	1.976	2.539	1.2	0.76	−0.01	−0.12
Martinique	0.472	–	2.44	–	–	–
St. Kitts and Nevis	0.11	0.141	1.7	0.83	−0.11	0.00
St. Lucia	0.311	0.404	1.32	1.02	0.14	0.21
St. Vincent and Grenadines	0.096	0.101	1.29	1.11	−0.18	−0.09
Trinidad and Tobago	0.557	0.464	0.11	0.27	0.43	−0.01
Turks and Caicos	0.292	0.706	3.4	1.37	−0.29	−0.14

if demand is adequate to recover the cost of investment. It is evident that small islands must meet the conditions proffered by Smith.

Besides tourism as a pathway may be riddled with growth pitfalls marring the development dream of small islands. In 2003, Smeral discussed the rising costs of tourism goods and services. According to his study, such cost increases could be explained if one accounts for the symbiotic relationship between two elements-the destination's production and the tourist's consumption of the tourism industry's goods and services. Moreover, these elements cannot be dissociated, as they are dependent upon the consumer's physical presence. For example, revenue that could be gained from sold hotel rooms is lost if the rooms remain unoccupied. Thus, without the consumer's presence, there is no impetus for producing the tourism industry's goods and services. As a result, the movement of opportunities that could flow and grow from the tourism industry become constrained. From Smeral's perspective then, a destination's tourism costs rise in that its productivity is more obstructed than that of a destination's manufacturing sector.

Due to its productivity gap compared to other economic activities, tourism development may reveal unintended consequences when applied in resource-constrained and vulnerable settings such as small islands. This possibility involves two aspects. The first aspect is related to the economy's size, so we must inquire if a small size or scale may hamper or induce growth. Mainstream economics is pessimistic about the economic prospects of small islands. The second aspect is related to investigating if tourism development is growth-enhancing. Again, mainstream economics seems skeptical about tourism development's economic merits due to its inherent productivity gap. I alluded to these studies in the previous chapters.

What is noticeable from the literature is that economic size and structure are factors in determining economic growth and prosperity. Examining these two factors is pertinent and meaningful since small islands struggle to come up with a structural answer that can lift them from the scale syndrome. It is critical to investigate whether small size affects economic growth, as Smith argued, because of new empirical evidence contesting Smith's view. New evidence suggests that the specialization pattern could better explain economic performance, indicating that on average tourism specialized countries grew faster than those nondependent on tourism.[25]

---

[25] See, for example, Looney (1989); Baldacchino and Milne (2000); Escaith (2001); Armstrong and Read (2001); Wint (2002); Jayawardena and Ramajeesinghh (2003); and Bertram (2004).

Tourism is low in R&D intensity, and, in general, tourism-dependent countries are small in size. So, why, and how are small tourism islands able to propel this growth despite the claims advocated in the extant economic literature whereupon smallness was a dooming quality?

5.5   TOURISM SPECIALIZATION (TS), SIZE, AND ECONOMIC GROWTH

If smallness of scale is a problem, can this limitation be overcome, and, if so, how? Some scholars, such as Arthur Lewis (1955), saw the opportunity of overcoming manufacturing scale problems through specialization, where trade can substitute for domestic production. William Demas (1965) saw the solution in aggregation through regionalism to create economies of scale. Others weighed in and saw the solution in breaking free from the capitalist world system.[26] Lewis's foresight about trade as a development device was limited to tradable services and industrialization (excluding tourism) as the "progressive" sector, but this turned out to be limited to small countries.

However, small islands specialize in tourism when empirical data analysis reveals the potential for positive outcomes. Table 5.2 compiles the TS of 23 small islands. These islands represent a geographic dispersion across the globe exposing differences in tourism life cycle, TS, and population. The most intense TS islands are Macao, followed by Maldives, and Aruba. The least intense TS islands are Bahrain, Cyprus, and Malta, St. Vincent and the Grenadines, and St. Kitts-Nevis. What is interesting from this table is that TS appears not to respect population size. For example, Maldives has a larger population than St. Kitts Nevis, while revealing a higher TS pattern.

What does the empirical evidence tell us when we compare the GDP per capita between small and larger population sizes? Table 5.3 shows descriptive statistics summarizing a comparison between 1995 and 2010 comparing the economic growth performance among countries and the world. Over this period, countries with a population of less than 1.5 million and a prominent presence of the tourism sector (share of tourism to GDP is greater than 20%) experienced a real per capita GDP growth of 2.35% per

---

[26] See, for example, Bryden (1973); Perez (1974); Turner (1976); and Britton (1982).

annum, and the world average was 1.7%.[27] The developed countries grew at 1.46% per annum. It is interesting to note that countries with fewer than 1.5 million inhabitants were mainly islands. Small islands with a TS rate exceeding 20% grew faster than the OECD countries. The countries with a population of less than one and one half million are both proliferating and increasingly specializing in tourism; they are small and open economies in the trade-theoretical sense.

While small island destinations with a TS exceeding 20% performed well compared to other regions, there were performance differences. For example, closer scrutiny of economic performance within the panel of countries in the Caribbean reveals an interesting variance pattern. Table 5.3 provides a view of this variance among some selected Caribbean countries. According to the World Travel and Tourism Council, Aruba, and St. Lucia are highly specialized tourism countries. They experienced a real GDP per capita growth from 1986 to 2010 of 3.8%, which was more than twice as high as the world growth rate and nearly 2.5 times higher than the OECD countries. Aruba excels in this group of countries.

This accomplishment is more remarkable if we consider the context of population growth. Aruba's population grew during the same time span about 3.5 times faster than the sample of countries, yet it generated more economic output productivity. Size, therefore, cannot be attributed as the determining factor in explaining the exposed variance. The degree of TS seems to have a significant positive bearing on the nexus between tourism and growth, according to Escaith (2001); Lanza et al. (2003); Croes (2005); and Sequeira and Nunes (2008).

As indicated previously, studies from Easterly and Kraay and Lanza and Pigliaru reveal that small islands exceed many larger countries in per capita GDP. The meaning of their studies reveals that, while small size may include disadvantages there are no significant differences when economic growth is compared between large and small countries. These studies corroborate the conclusion of the Armstrong et al. (1998) study, which also suggests that size does not account for the per capita GDP level of small countries. These findings and the results depicted in Table 5.3 question the validity of the assumptions of the neoclassical view of development economics and development. Hernandez-Marin posits that:

---

[27] In Chapter 2 we found a correlation between country size and openness. That is why this table followed the small island literature convention to combine openness (tourism specialization) and an island's population.

TABLE 5.2   23 Countries with the Highest Tourism Specialization

Country	Tourism Receipts/GDP	Population	Small	Island
Macao	0.81	536,969	Yes	Yes
Maldives	0.66	364,511	Yes	Yes
Aruba	0.51	101,669	Yes	Yes
Palau	0.42	20,470	Yes	Yes
Seychelles	0.36	89,770	Yes	Yes
Vanuatu	0.35	236,295	Yes	Yes
Curacao	0.325	148,703	Yes	Yes
Fiji	0.26	859,950	Yes	Yes
Antigua and Barbuda	0.26	94,661	Yes	Yes
Barbados	0.24	279,569	Yes	Yes
St. Lucia	0.23	172,580	Yes	Yes
The Bahamas	0.23	360,832	Yes	Yes
Cape Verde	0.23	502,384	Yes	Yes
Dominica	0.19	71,440	Yes	Yes
Samoa	0.19	186,205	Yes	Yes
Jamaica	0.17	2,817,210	No	Yes
Mauritius	0.16	1,250,400	Yes	Yes
Grenada	0.15	104,677	Yes	Yes
St. Kitts Nevis	0.13	51,445	Yes	Yes
St. Vincent and the Grenadines	0.13	109,315	Yes	Yes
Malta	0.13	414,508	Yes	Yes
Cyprus	0.1	1,112,607	Yes	Yes
Bahrain	0.08	1,240,862	Yes	Yes

Source: World Development Indicators, World Bank. http://data.worldbank.org/data-catalog/world-development-indicators.

TABLE 5.3   Economic Growth of Country Groups, 1995–2010

Country Group	Growth of GDP per capita (%)
OECD	1.46
High income non-OECD	1.52
Small	2.86
Small tourism >20	2.35
Small tourism <20	2.88
World	1.70

Source: World Development Indicators, World Bank, http://data.worldbank.org/data-catalog/world-development-indicators. And author's own calculation. The group of small countries with tourism exceeding 20% of their GDP consists of Aruba, Antigua and Barbuda, The Bahamas, Barbados, Cape Verde, Fiji, St. Lucia, Maldives, Palau, Seychelles, and Vanuatu.

"The aforementioned works seem to support the hypothesis that size is not a sufficiently important determining factor in economic development, and that, in any case, there are other more relevant variables which prevent the relationship between the size of a country and its economic results from being clear cut" (2008, p. 2).

This is a comforting and encouraging conclusion because it shows that success is possible despite the small size. In other words, size does not determine the validity of the TLG hypothesis; other conditions seem to be at work. However, it is not clear whether TS will spawn growth on a sustainable basis or whether the positive relationship may trigger a diminishing growth rate as TS deepens. Table 5.3 reveals that there may be other influences at work, mediating the impact of TS on economic growth. Arguably, small island destinations with less than 20% TS have performed better than those destinations with a TS exceeding 20%. Also, the positive relationship between tourism and economic growth may also hinge on the use of foreign receipts. Will these receipts be used to pay for capital and human capital improvements, or will foreign receipts be used for merely consumption purposes? These examples would fit into the tourism area life cycle (TALC) perspective. TALC seems to imply that destinations have a life cycle that inevitably would lead to the decline of a destination.[28] Some small islands have experienced a decline in the past, such as Curacao in the Caribbean and the Balearic Islands in the Mediterranean. There seems corroboration in the tourism literature to support the non-linearity destination evaluation following the TALC model.[29]

5.6 OPENNESS AND ECONOMIC PERFORMANCE

Some scholars have argued that a country's openness to the global market is an important explanatory variable.[30] The idea is that the greater openness

---

[28] For a discussion on the TALC framework, see, for example, Cole, S., (2007). Beyond the resort life cycle: The micro-dynamics of destination tourism. *The Journal of Regional Analysis and Policy*, *37*(3), 266–278. See also, Weaver, D., (2005). The "plantation" variant of the TALC in the small island Caribbean. In: Butler, R., (ed.), *Tourism Area Life Cycle: Applications and Modifications* (Vol. 1, pp. 185–197). Clevedon, UK, Channel View Publication.

[29] The question is whether the positive relationship between tourism specialization and economic growth can last? There are some indications in the literature that suggest that the positive relationship is not constant and maybe even decreasing in time. See, for example, Amadou, A., & Clerides, S., (2010). Prospects and limits of tourism-led growth: The international evidence. *Review of Economic Analysis*, *3*, 287–303. We will discuss this prospect in our next chapter.

[30] See, for example, Sachs and Warner (1995); Frankel and Romer (1999); and Easterly and Kraay (2001).

to international trade, the greater the likelihood of growth. Trade theory posits that nations export to others based on comparative advantage, and those nations will improve economically in the process. Openness through exports has a positive impact on the economic output of a country. Specifically, international trade is critical to growth for small countries because it mitigates the constraints imposed by the small domestic market, increases efficiency through greater competition, and provides critical financing for essential imports. This theory maintains that exports promote growth through several channels, such as economies of scale, the potential of positive externalities in non-export sectors, the encouragement of efficient allocation of resources through increasing competition, the stimulation in R&D investment and human capital, and the loosening of foreign exchange constraints.

The greater level of openness-as defined by the ratio of the imports and exports to GDP-is perceived as a determining factor in enhancing national economic output. The empirical evidence on the relationship between export and growth, however, is mixed. This empirical evidence is at the heart of why some nations grow more than others. These mixed results are the consequence determined from three different theoretical sources. These theoretical frameworks shifted from the resource-based approach, the endogenous approach embedded in technological change to openness or trade as economic growth sources. We already alluded to Kaldor, who characterized growth as export-led. An important issue in this debate is whether or not a country's specialization pattern has a bearing on growth.[31]

I contend that growth rate characteristics and differences can uncover variations occurring in the elasticities of demand. Moreover, in the context of ToT and demand growth lay the specialization patterns that can be characterized as favorable. Indeed, it is this context that appears familiar in the tourism development landscape. For example, small islands translate their non-tradable goods into marketable entities. Therefore, tourism forms and extracts positive links to the world economies. It may be debatable whether small islands design and develop a tourism industry based on their comparative advantage over competing markets. According to trade theory, such advantage exists under circumstances of high productivity

---

[31] For a debate, see Barro (1991); Sachs and Warner (1995); Frankel and Romer (1999); and Jin (2004), among others, who attest to the positive results paradigm. Harrison and Hanson (1999) and O'Rourke (2000) questioned the robust positive relationships between these two variables. The debate highlights two flaws, which might have caused this confusion, namely methodology and estimation techniques on the one hand, and issues of aggregation of data (Jin, 2004; and Durbarry, 2004).

or because the tourism destination possesses multifaceted resources that support and contribute to tourism as a viable means toward economic prosperity. Chapter 2 already discussed the implications of trade theory, where I posit that the dynamic comparative advantage anchored in the division of labor of Adam Smith is better suited to explain the choice for TS. I argued that surplus production capacity and smallness are the source of the TS choice. Again, smallness can only be conquered through international trade.

The declining agricultural sector and the phasing out of preferential trade agreements that supported the declining economic sectors indeed provoked a surplus capacity production in many small islands. A case in point is the so-called "Banana Trade War." This trade war engulfed St. Lucia and St. Vincent's agricultural sectors, which ended in January 2006 when the preferential access to the European Union (EU) banana market ended for these two islands. Needless to say, that the costs for these two islands were steep: unemployment skyrocketed, and revenues declined steeply.[32] A similar situation happened when Fiji lost its preferential access to the EU sugar market in 2005. The surplus capacity production reveals itself in among others, high unemployment, and substantial migration (MI).[33] For example, when Aruba decided to industrialize tourism, it was because of the high unemployment and the surplus production capacity. The unemployment in Aruba in 1985 clocked nearly 25% of the working population due to the failed oil industry. High unemployment was also the hallmark of many Pacific islands, which is a consequence of idle production capacity.

Economic growth and development in small islands have long been attributed to trade. Indeed, the outcome of trade has registered extensive growth measures beyond that of even larger countries. As noted in Chapter 4, some studies, including that of Helpman, attributed growth differentials between small and large countries to the degree of openness. According to Helpman, "just one standard deviation in the degree of openness positively impacts small country growth rates by more than three times that of a larger country."[34] What is significant here is that openness can

---

[32] See, for example, an interesting account of the banana trade war in Barfield (2003); Payne (2006); and Bishop (2010).

[33] Regarding the substantial migration in the South Pacific and the Caribbean, see, for example, Connell and Conway (2000).

[34] See, for example, Helpman (2004) assessed the effects of trade on economic growth comparing Mali and Seychelles.

balance the vulnerability that small countries experience and suffer in the global market. Thus, whereas tourism industry countries such as Aruba, Barbados, Malta, and the Maldives necessarily participate in the global market, they still indicate confirmed and conclusive evidence of growth-doing so over time.

There is, then, given this example, testimony that the degree of openness can boast convincing, positive sway on growth rates. Table 5.4 suggests that countries with a high degree of openness also reveal higher affluence. There seems to be, however, a less consistent correlation between openness and the levels of income of the countries under review. For example, Haiti reveals a similar degree of openness as Jamaica in 2010. Yet, Jamaica's GDP per capita was more than seven times compared with Haiti's GDP per capita (Table 5.4).

5.7   SPECIALIZATION, LINKAGES, AND PRODUCTIVITY GAP

Kaldor suggested that specialization could be the manner by which tourism could develop a path to economic growth. Kaldor's claim aligns with the proposition that what a small island specializes in matters for growth. In keeping with this literature, I examined the descriptive statistics that referred to TS, size of the country, and positive economic growth. This examination consisted of the annual data of 29 countries during the period of 1995–2017 via tourism receipts and documented economic growth. TS was greater than 10% in all 29 countries. In keeping with the literature, we define TS as the share of tourism receipts in GDP.

The scatter plot in Figure 5.1 demonstrates that TS yields a positive bearing on growth. Destinations that indicated less than 10% specialization yielded a growth rate that paled beside those that indicated greater than 10%. Figure 5.1 shows that the GDP per capita data (x-axis) and tourism receipts (y-axis) are charted. The points do indicate that there exists a direct and positive relationship between TS and the economic growth that results from that synergistic nexus. Also, note that islands tend to cluster together, revealing a sense that islands tend to have a similar development pattern. A breadth of economic sectors benefits from tourism at small island destinations. As tourism systems and processes are realized, spill-over advantages are harvested by related economic sectors such as transportation, banking, utilities, food, and beverage, and construction.

TABLE 5.4 Selected Caribbean Countries' Variance in Area, GDP, Openness, and Population

Country	Nominal GDP per capita		Population		Real GDP per capita		Degree Openness (%)		Area in sq. km 1,000
	2003	2010	2003	2010	2003	2010	2003	2010	
Aruba	20,835	24,272	97,627	101,669	26,557	23,469	133.96	136.32	0.19
The Bahamas	28,093	27,979	310,304	360,832	31,014	27,979	71.06	78.68	13.88
Barbados	12,029	15,959	269,418	271,440	14,984	15,959	88.30	88.25	0.43
Dominica	4,995	6,912	69,200	71,440	5,855	6,912	78.76	89.13	0.75
Dominican Republic	2,413	5,454	8,615,910	9,898,000	4,052	5,454	84.45	56.00	48.73
Haiti	330	662	8,217,710	9,993,000	724	662	63.75	80.09	27.75
Jamaica	3,465	4,683	2,614,430	2,701,000	4,875	4,683	88.15	80.93	10.99
Netherlands Antilles	16,552	–	175,480	–	14,558	–	–	–	0.96
St. Lucia	7,385	8,008	160,620	172,580	7,122	8,008	108.01	112.43	0.62
Trinidad and Tobago	21,188	16,684	1,298,280	1,328,000	12,426	16,684	89.80	85.76	5.13

Source: IMF, World Bank, UN Statistics, WTO, the Global Economy.com, and Central Banks of several countries in the sample.

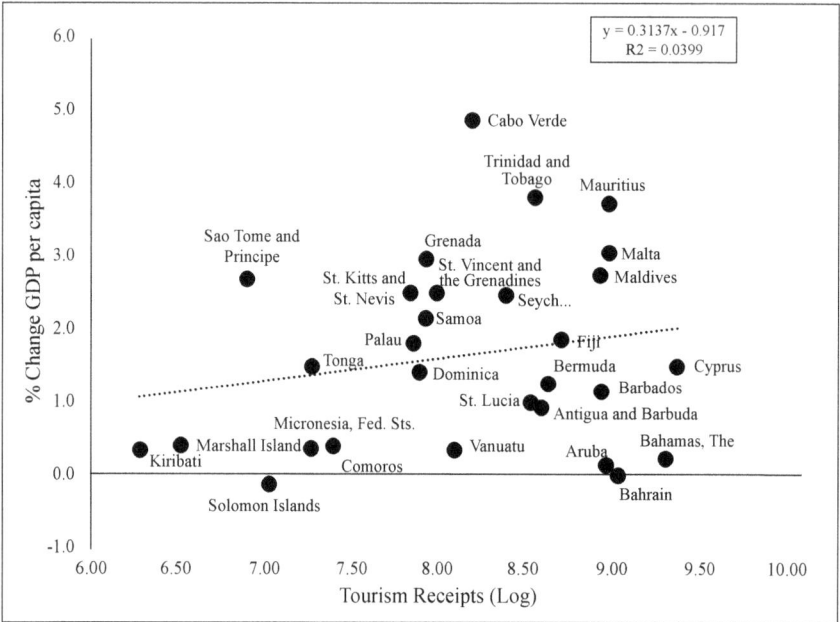

FIGURE 5.1   Per capita GDP growth in 29 small island destinations, 1995–2017.

Tourism generates backward and forward linkages, which reveal the interdependency level and degree in the economy. Backward linkages refer to the situation when the tourist payments in a hotel or restaurant beget payments in other areas, such as hotels buying from utilities, business services, and manufacturing. These purchases involve the completed hotel sales and generate the tourism multiplier effect in the economy. This economic multiplier effect varies between islands and over time. In a study of seven islands covering the Caribbean, the Indian Ocean, and the Pacific, Pratt applied a computer general equilibrium (CGE) modeling to estimate tourism's economic multiplier.[35] The results reveal that the average output multiplier of these islands was 1.49, and ranges from 1.06 in Fiji to 2.11 in the Seychelles. The income, on the other hand, had an average of 0.69, ranging from 0.49 for the Seychelles and 0.91 for Jamaica (see Table 5.5). The difference between the output multiplier (sales) and the income multiplier (the return of labor and capital that stay on the island) is the

---

[35] See Pratt (2015).

leakage that goes to the outside world and to the local government through taxation.

Island tourism activities also generate forward linkages. The accommodation sector, because of its business nature (tourist money is final demand: that is why the tourist money is the economic direct contribution), does not supply goods and services to other sectors. However, the transportation sector shows strong forward linkages in several of these islands. The interdependency generated by tourism to other economic sectors facilitates a quick learning opportunity of the tastes and preferences of tourists visiting the islands. Established or new local entrepreneurs quickly learn that the presence of tourists opens new opportunities for sales once they have ascertained the wants and needs of the tourists, including the cultural tourist.[36] Along with the established entrepreneur, newly incentivized entrepreneurs marshal in new ideas and products that meet tourist desires.

TABLE 5.5   Economic Contribution of $1 Million Tourist Expenditure

	Output	Income	Tourism Output Multiplier	Tourism Income Multiplier
America Samoa	$1,030,485	$439,876	1.03	0.44
Aruba	$1,433,338	$680,443	1.43	0.68
Fiji	$1,045,458	$913,565	1.05	0.91
Jamaica	$1,384,293	$915,022	1.38	0.92
Maldives	$1,710,980	$644,595	1.71	0.64
Mauritius	$1,362,653	$756,239	1.36	0.76
Seychelles	$1,814,990	$478,201	1.81	0.48

Source: Adapted from: Pratt (2015).

Yet, the literature does not appear to provide decisive support regarding this disseminated information. That is, for example, Jackman, and Lorde in 2010 determined that with regard to Barbados TLG was the result of leakages, which encouraged higher imports as opposed to TS-thus relegating the positive consequence of TS on economic growth inconclusive.

Pratt's findings do not mesh with typical critiques fielded against tourism development. Some commentators connect the leakage problem

---

[36] Tourists' direct spending is not as germane to local vendors as the effects of linkages spawned by the tourism industry in supply chains. See, for example, Lejarraja and Walkenhorst (2010).

with mass tourism that is typically associated with small islands.[37] Leakages include monies disbursed for tourist expenditures. Such expenditures are considered leakages and are dispensed as wages paid to external entities beyond the destination, taxes (including those that have been repatriated), and any goods and services that had been imported. External leakages account for monies allotted for travel agents, cruises, booking prices arranged as a consequence of, e.g., special offers, etc. Such external leakages can also be disbursed to foreign investors who provided financing for necessary tourism structures. However, other empirical studies are consistent with Pratt's results. For example, Pratt's results regarding Aruba are very similar to the results of the Steenge and van de Steeg study.[38]

The latter study found an income multiplier of 0.685 for the year 1999. However, the study suggests when including the induced effects, the value of the income multiplier increased to 0.952. The leakage factor, based on these results, would be 0.448. The latter result is similar to our own estimate, which is 0.438. Our estimate is an average covering data from 2010 to 2017 (see Annex 2) Tables 5.6 and 5.7 reveals our results. The leakage estimates of Steenge and van de Steeg, based on 1999 data was 0.448, which is consistent with our own estimate.[39] These findings are higher than those reported by the World Travel and Tourism Council, which reveals that leakage could rise to 30–40% in the Caribbean. Although the leakages are significant, the tourism industry continues to maintain its economic significance to these small island countries.

The ambiguity between TS and economic growth may stem from the tourism productivity gap. Tourism is not strong in productivity gains, as indicated in the previous chapters.[40] It lags in productivity compared to manufacturing, which affects its production cost's structure. As production costs tend to increase, we would argue that tourism's positive relationship with economic growth is in charging higher prices for its offerings. However, do higher prices depress demand? It depends. We notice that the destinations enjoying the highest tourist spending are also those that charge higher prices for their offerings. For example, destinations in Western Europe such as France, support this contention. This implies

---

[37] For example, see Scheyvens (2002); and Bishop (2010).
[38] See Steenge and van der Steeg (2010).
[39] Pratt's data was from 2010 and his leakage was higher because it also included taxation paid to the local government. I am indebted to Jorge Ridderstaat regarding these comparisons and estimations.
[40] See also, for example, Keller and Bieger (2007).

that the relative price competitiveness of destinations may spur demand. This price differential may be the outcome of management efficiency and creativity that ignites value for the customer.

Price competitiveness differences in the Caribbean do not seem pronounced. A selective analysis of price competitiveness in the Caribbean by the World Travel and Tourism Council in 2009 reveal that differences are modest, revealing about a one-point difference considering the Dominican Republic with an index of 4.5 (out of 7 scale score) to a 5.5 index in Trinidad and Tobago, with Barbados in the middle with an index of 4.4. By 2017, price competitiveness deteriorated for Barbados to 3.0 and Trinidad and Tobago (4.8), while the Dominican Republic (4.4). However, price competitiveness does not seem to shape arrivals to each destination. According to these perspectives, TS results from an abundance of resources (natural and manmade attractions) or price differentials to develop the supply of unique tourism products.

Arguably, small islands cannot grow forever based on more tourist arrivals due to supply constraints (limited land availability). One could imagine a situation where a small island may deplete its natural resources by bumping into its landmass boundaries. For example, a small island of 130 square miles cannot build 40,000 hotel rooms due to land constraints and potential residents' hostility. The maximum number of tourists that they can welcome is bounded, which means they must limit the number of visitors to preserve future profits.[41] This assertion has already been a major concern in some island destinations in the Mediterranean and the Caribbean. For example, as discussed in previous chapters, the Balearic Islands were successful in upgrading their tourist product and perpetuating the SSS product. Aruba also faced the question of "how far and how fast" its tourism industry could or should expand in order not to affect the positive relationship between TS and quality of life.[42]

Kaldor's pathway to economic growth premised on functional exports defined by the elasticities of demand in source countries empirically appears to be a feasible pathway for small islands. I shall empirically investigate this question in the next chapter.

---

[41] For an excellent discussion of trade-offs between arrivals and natural resources, and carrying capacity, see Marsiglio (2018).
[42] See, for example, Cole and Razak (2009).

KEYWORDS

- demonstration
- development strategy
- economic growth
- economic specialization
- gross domestic product
- tourism-led growth hypothesis

CHAPTER 6

CONSTITUTIONAL ECONOMICS: BLASÉ?

In *Lessons from the Political Economy of Small Islands*, Baldacchino and Milne referenced the relevance of jurisdiction as an economic resource.[1] In their view, jurisdiction means the smart deployment of legal and political resources to achieve economic growth and prosperity goals. The implication is that sovereignty or political independence for small islands may not be the only or smartest choice in the decolonization process. Prosperity is more critical than power, according to this view. In their words:

> *"...what is important may not be the pursuit of some ultimate model of political sovereignty, which may well be inappropriate in a particular context, but rather determining what kind of constitutional or organizational arrangements can provide the necessary local power for the achievement of a community's dignity and prosperity." (p. 10)*

Their study's premise is that jurisdiction is an essential component in triggering and shaping economic capacity, and if used strategically, may help small islands mitigate against and overcome size constraints. This perspective recognizes the constraints imposed by small size, such as a restrained domestic market and limited natural resources. It claims that these constraints can be overcome or mitigated because small islands may use their jurisdiction to draw resources and assets from beyond their boundaries, through either high openness or integration with the outside world. Baldacchino and Milne (2000) called the jurisdictional advantage the "economics of constitutionalism." This view claims that a legal jurisdiction short of independence is the explanatory variable of growth and prosperity.

---

[1] See Baldacchino and Milne (2000).

These ideas are not new. Small islands have played this jurisdiction card as an economic resource in multiple settings and platforms, coming up with creative solutions to offset scale disadvantages. Two examples illustrate the potential of jurisdiction as an economic resource. The Cayman Islands, a three-island dependency in the Caribbean, is a global financial haven. The Cayman Islands used its jurisdiction as an economic resource to provide a means of sheltering offshore financial assets. The Caymans have maneuvered and established itself as a financial sanctuary for investors, international corporations, and various businesses where such actors benefit from the advantageous legislation that, by its Constitution, has abolished business taxes such as corporate and income. Thus, the Cayman Islands function as a tax haven-unlike other global countries where taxes on anything from capital gains to gift taxes are a burgeoning slice of profit. The tax exemptions function as a magnetic body affording island-based or offshore enterprises their desired profitable protections.

The Aland Islands claimed dependency from Finland in 1921.[2] However, the islands extracted a home rule that guaranteed them the right to keep their identity (Swedish language and traditions) while obtaining large self-government. The constitutional arrangement with Finland protected the Aland Islands from any international consequences stemming from Finland's decisions without the acquiescence of the islands. For example, Finland's decision to enter the European Union (EU) in 1995 would have no effects on the Aland Islands without their confirmation. Because of this special jurisdiction, the islands extracted far-reaching exemptions from EU regulations. Moreover, the Aland Islands remained outside of the EU tax regime. The islands turned this special jurisdiction with Finland and the EU into an economic asset nurturing a comparative advantage through specialization in maritime transport and related services.[3]

Small islands have often attempted to enlarge their markets by special arrangements with previous metropole governments, such as preferential access to their agricultural products, bilateral, and multilateral aid. Regarding trade preferences, there are small islands such as islands in the Caribbean that have been successful in getting market access to the United States and Europe under beneficial conditions. For example, under the African Caribbean Pacific (ACP) regime, the EU granted preferential

---

[2] The Aland Islands is a Swedish-speaking archipelago in the Baltic Sea belonging to Finland and has about 30,000 inhabitants.
[3] See, for example, Lindstrom (2000).

market access to small islands' products, including garments, sugar, rum, tuna petroleum, banana, and fish. Similarly, under the Caribbean Basin Initiative (CBI), the United States conceded preferential treatment to Caribbean products, including methanol, polystyrene, steel products, electrical products, ethyl alcohol, pneumatic tires, rum, and yams.

The political instability and economic uncertainty of the Caribbean of the latter 1970s and into the 1980s urged the United States Congress to create the CBI. With such political and economic issues, the tourism industry, in its importance to employment opportunities and a means to gain foreign exchange earnings (Caribbean Basin Economic Recovery Act, 1989), was recognized by the United States as vital to its welfare. Accordingly, in 1984 with amendments in 1990 and 2002, the United States Congress, through legislation and via the CBI, authorized, and empowered the Caribbean islands to procure advantageous trade status. Further, the CBI augmented aid to bolster the Caribbean's infrastructure while incapacitating some trade barriers between the United States and the Caribbean.

Overall, Americans could deduct business expenses associated with their attendance in business meetings and conventions-provided such countries (such as the Caribbean) holding the events maintained a CBI status. Moreover, such a trade concession allowed attendees to circumvent the kinds of restricted deductions compelled by other global entities (Caribbean Basin Economic Recovery Act, 1989). Albeit, CBI countries could not create or maintain tax laws that, in effect, could victimize conventions that were held in the US (International Trade Administration, 2000).[4] Small islands such as Mauritius, Trinidad and Tobago, the Bahamas, Seychelles, St. Lucia, Barbados, and St. Kitts Nevis have taken advantage of these preferential programs.[5]

Small islands also embraced international tourism as a way to increase the domestic market through international tourism demand. At the outset of tourism promotion as an economic strategy, *Lady Fortuna* embraced the Caribbean Islands when the United States imposed travel restrictions to Cuba in 1963. The latter meant a *de facto* protection of the tourism industry in other Caribbean islands. This protection augured well for

---

[4] Croes and Schmidt (2007) suggested reigniting the travel benefit provision of the CBI law. They saw this provision as a demand-pull approach based on market incentives to support travel from the United States to the Caribbean, rather than continue to rely on existing flaws in foreign aid delivery.

[5] There is an ongoing discussion on whether small islands should have special treatment because of potential vulnerabilities induced by their small size. See, for example, Hein (2004).

regional tourism because tourism development flourished on these islands. We already discussed the impact of an eventual lifting of US travel restrictions to Cuba on Caribbean islands in a previous chapter. In the Mediterranean, Malta also embraced tourism as a way to deflect economic and social hardships.

Small islands also spurred affluence through remittances from "transnational corporations of kin." The special relationship with metropole countries allowed islanders to migrate to these countries, searching for work and opportunities. As migrants sought employment and beneficial opportunities beyond their borders, they contributed to a multifarious blueprint of movement that characterized an affiliation and convergence between their region and the global community. The resulting enters and exit of this movement throughout ensuing generations served to create and enhance a labyrinthine economic and environmental accord between regions of origin and the international stage. By way of remittances and the human capital that migrants bring back to their home region exists a social and cultural benefit that strengthens the regional community.[6]

Not all island states were able to draw advantageous affiliations. Many small island states achieved independence by default simply because there were no apparent partners with whom to develop a political alliance, as in Sao Tome and Principe's cases. After decolonization of the 1970s, many new island states pursued autonomy and recognition, which was fueled by the rediscovery of their historical roots and calculations of comparative advantage-perhaps inspired by a deep sense of identity, perceived discrimination, neglect, exploitation, or a history of repression. Thus, many small island states may serve as obvious examples of resurgent localism. Indeed, the 20th century has been the century of the triumph of self-determination as small island states worked to identify and direct their destinies.

Small islands have stubbornly refused to accept the will of larger countries. They have either obtained political sovereignty or maintained their non-sovereign status. Pundit predictions indicated that several small islands would become independent in the 1990s: Montserrat, Aruba, and the French overseas departments.[7] These predictions turned out to be incorrect. One island, Aruba, changed the Charter of the Dutch Kingdom to

---

[6] The concept of remittances is significant as their impact on island economies not only imposes a central necessity but could also bear a warping of the economics. Given their economic and environmental vulnerabilities, small island microstates are perceived as MIRAB economies beyond the Pacific region. See, for example, Bertram (1986); and Connell and Conway (2000).

[7] See, for example, Sutton (1987).

release itself from the obligation to become politically sovereign in 1996. Fueled by sentiments of self-identity and the calculations of comparative advantage, Aruba has held steadfastly to the principle that jurisdiction is a strategic, crucial resource for growth and prosperity. By empowering itself with the constitutional capacity to act for itself in economic strategy and planning and the freedom to use its powers more intelligently, Aruba believed it could build a more self-reliant and diversified economy while providing a better quality of life for its peoples. Self-determination in this context was seen not only as a validation of its identity but also as a tool for controlling its own economic space and for seizing niche opportunities in a global economy. Aruba's case may still have similarities regarding the strife other countries endure toward independence.[8] Aruba's vision was epitomized in pursuing self-government without financial dependency, an ideology that came to be known as *Status Aparte* or Separate Status.

6.1 THE RHEOSTATIC LEVER OF CONSTITUTIONAL ECONOMICS

Some 20 years ago, I was present at the University of Aruba auditorium, where Gert Oostindie, a professor at the University of Leiden, introduced his book about the modern constitutional history of the Kingdom of the Netherlands.[9] The presentation happened in a formal setting in the auditorium, which is a 19th-century small chapel built around 1937 to serve the religious needs of the Friars of Tilburg.[10] I did not know what to expect of or from the presentation, or if the presentation would recognize the Aruban people's efforts to obtain and retain their much-cherished *Status Aparte*. I sat to listen with mixed feelings and some trepidation about what bastion the presentation would herald. I hoped that the Dutch account of Aruba's constitutional story would reflect the determination and inspired agency of the Aruban people to force The Hague to acquiesce to the will of Aruba's

---

[8] Aruba's fortune was that it was tied to two levels of control (a strict one (Curacao) and a more flexible one (The Netherlands). It got away from the first one, which allowed for a large degree of self-determination, enough to turn around its economic structure and reap the benefits from this. In a sense, Aruba still underwent the effects of "economics of constitutionalism," though not in the islands' full sense that got their independence.

[9] Oostindie, G., & Klinkers, I., (2001). *Knellende Koninkrijksbanden. Het Nederlandse Dekolonisatiebeleid in de Caraiben, 1940–2000.* Amsterdam: Amsterdam University Press.

[10] The friars were housed in the Huize van La Salle; the building transferred to the University of Aruba in 1988.

people. On the other hand, it was also possible that the Dutch account would be about portraying Aruba as nothing more than building castles in the air. Well, there it was. The presentation recognized nothing. I also expected that the Dutch account would portray the Aruba story as a reflex of the Dutch will. So, there it was, and there I was, sitting, and listening in dismay. The presentation recognized nothing of the Aruban people's intensity and commitment to eschew independence as the outcome of their expression of self-determination.

Oostindie's book provided a glimpse into the history of the Kingdom of the Netherlands and the political intricacies that shaped the constitutional outcomes since the Second World War. What surprised me about his study was that of the four partners, i.e., the Netherlands, Surinam, the Netherlands Antilles, and Aruba, only Aruba has triumphed in wills since the establishment of the Charter of the Kingdom 50 years ago. Suriname became independent against its will, the Netherlands Antilles disintegrated against Curacao's will, and Aruba gained its *Status Aparte* in 1986 and maintained this special status despite the resistance of The Hague and Willemstad. Aruba was supposed to become independent in 1996, according to the Charter of the Kingdom.[11] The presentation of Oostindie was strikingly devoid of recognition of the active agency of the Arubans to achieve the political status they desired-even against all the odds. The presentation felt suffused of the toxicity of ignoring the political vision and the foundations of that vision to stay within the Dutch Kingdom for instrumental reasons, and the reveling of self-congratulatory eulogies to the Dutch humanist approach to the islands.

How did Aruba, the smallest of the four entities, realize its will despite the enormous disparity in power with the Netherlands? The classical theories of power would not have predicted this outcome. They would have predicted that Aruba would have failed in such a battle and be forever relegated to a beggar town's peripheral status, more befitting its small size. This negative connotation of small size is premised on the neoclassical framework discussed in previous chapters, which traditionally has equated small size with insufficiency: population, labor, diseconomies of scale, entrepreneurship, and capital of resources and intellectually creative

---

[11] Overcoming the will of the mother country is not solely associated with Aruba's constitutional experience. In 1967, Anguilla experienced a similar experience where the will of Anguilla superseded that of Whitehall. The British colonial office wanted to force Anguilla to join the union of St. Kitts and Nevis. After some widespread upheaval on the island, the British created a separate dependency to accommodate Anguilla and St. Kitts and Nevis became independent.

properties in general. According to this view, the insufficiency of resources affects the performance of these entities in such a way that they cannot achieve convergence with more developed countries. The high openness and integration with the global market-a hallmark of small countries-has been perceived as a sign of dependency and vulnerability, which might hurt growth performance. This view clashes with Alesina and Spolaore's claims, espoused in Chapter 2, about the opportunities that openness assumes for small countries.

The 50-year debate in the Dutch Kingdom about respecting the people's will through their right of self-determination was also clouded by this neoclassical framework, which equated smallness with weakness. The discussion of exercising the right of self-determination of the Aruban people was confined from the 1940s through the 1980s to only one possible outcome: independence. Because independence was the only possible result of exercising that right (it was only natural that the discussion's premise would hinge upon the viability of Aruba). The discussion was whether Aruba's independence could be self-sustaining in the broader global setting. The conventional wisdom maintained that a small island such as Aruba could not overcome its small size with all of the constraining inflictions that small size compels and levies against itself. So, it could not become independent and, hence, the right to self-determination of the people should be confined to a context of their allegiance to a larger sized authority. The entire debate within the Dutch Kingdom centered for decades on only two alternatives: independent or dependent status within the Netherlands Antilles as a federation.

Pursuing the constitutional levers for prosperity was necessary, given the rapidly changing conditions in the global world. Aruba would need to survey flexible options due to these changing realities if it was to find its road to prosperity. In the 1980s, Aruba's determination was tested severely when it lost its central economic pillar and significant economic development source-oil refining. This harsh reality underscored the argument that survival, prosperity, and sustainability could only be accomplished through an appropriate legal infrastructure that could respond quickly to such changes. Aruba realized this legal infrastructure could never be accomplished within the constellation of the Netherlands Antilles.[12]

---

[12] There are several accounts of the constitutional history of Aruba. See, for example, Croes (2011).

Aruba's ambitions were not well received in The Hague and Willemstad. The centers of power for the Dutch Caribbean accepted the conventional wisdom that smallness inhibits growth performance. They insisted on denying Aruba's right to self-determination; however, eventually, Aruba obtained and retained its *Status Aparte* within the Kingdom of the Netherlands. When Aruba assured its *Status Aparte* while remaining part of the Kingdom of the Netherlands, The Hague's governing power was grossly reduced. No longer would Aruba accept or allow The Hague's authority to hold sway over finance complexities such as taxes, government budgets, development, education, social, and cultural issues, and concerns. Nor would it impose its power over political commissions and actions such as immigration. Infrastructural undertakings such as transportation would remain the sole responsibility of Aruba. In short, Aruba would assume power over and governance of its destiny under its people's wants and needs. The Hague would maintain responsibility for military protection, foreign affairs, and the processes affording and controlling citizenship.

Empirical evidence suggests that non-independent islands have fared much better than politically independent islands. The premise is that bonding constitutionally with the mother country seems to provide more benefits to the islands rather than independence from the mother country. Several reasons support this contention. The non-independent island can prosper from the foreign aid proffered by the mother country and the necessary legal security to promote business transactions and economic growth. Besides, enjoying the same citizenship as residents in the mother country provides ease of travel and mobility when needed and access to first-class higher education, aid-financed infrastructure, and communications, natural disaster relief, provision of costly defense, and access to external capital.[13] These advantages made islands push to have a greater integration with the mother country, the very antithesis of sovereignty. Such is the case of the Dutch islands Bonaire, Saba, and Statia.[14] According to Baldacchino, "…the smaller, mainly island, colonies have either obtained political independence last, or else continue to stubbornly refuse to shift

---

[13] See, for example, McElroy and Mahoney (1999). For example, Montserrat received much-needed disaster relief from the British government during 1995–1997 after the volcano Soufriere eruption.

[14] On October 10, 2010, the Netherlands was dissolved and split into two constituent countries within the Kingdom (Curacao and St. Maarten) along Aruba's lines and three wholly integrated islands (Bonaire Saba and Statia) as special municipalities into the Dutch administrative body. This constitutional action came to be known as 10-10-10 (standing for October 10, 2010). For discussing this constitutional episode in the Netherlands Antilles, see, for example, Goede (2016); and Veenendaal (2015).

gear from their current, non-sovereign status." [15] By Baldacchino's count, at least 41 island cases have been revealing this latter position.

Empirical studies also propagated support for the claim that small islands are better off not being politically independent. McElroy is perhaps the contributor who has engaged in this comparative analysis more than any other commentator. In several studies, he concluded that small, dependent islands have fared much better in restructuring their economies and increasing their prosperity. In a study of McElroy and Mahoney comparing dependent countries (comprising territories freely associated or politically integrated with metropolitan nations) and independent states, they found that dependent islands in the Caribbean had per capita GDP of $11,214 compared to $5,898 for sovereign independent island states.[16] McElroy and Sanborn also asserted that residents of dependent islands compared to the independent have higher per capita income and electricity consumption, are more literate, experience much lower infant mortality, have better health care, and live longer.[17]

Another study of McElroy concluded that political dependencies are sources of growth and opportunity with growing employment and economic momentum.[18] Working with a larger global sample of small economies, Armstrong, and Read also reported a significant negative association between political sovereignty and per capita income. This idea that non-independent islands are more likely to prosper economically resonated with other analysts. For example, Poirine showed that non-independent islands did economically better than independent islands. Dunn also suggests that political dependence benefits small islands.[19] Furthermore, Rivera also pointed out that "The perception that the non-independent countries' economies are more 'modern' and prosperous than those of most of the independent countries…" is correct.[20]

---

[15] Several accounts of how small islands have resisted independence either attaining independence as the last remnants of colonization or have assumed after aggressive political pressure on the mother country, a legal and political jurisdiction short of political independence. Islands such as Aruba, Bermuda, or Reunion remained dependencies despite prognostications to the contrary. See, for example, Baldacchino (2006).

[16] Arguably, not every small dependent island did better than small independent islands. Several examples of islands, such as Montserrat, underperform islands such as Barbados and the Bahamas. See, for example, McElroy and Mahoney (2000).

[17] See McElroy and Sanborn (2005).

[18] See McElroy, J., (2006). Small island tourist economies across the life cycle. *Asia Pacific Viewpoint*, *47*(1), 61–77.

[19] See Poirine (1998). Also, see Bertram (2004).

[20] Cited in McElroy and Sanborn (2005).

I have visited many Caribbean islands and have seen the steep differences between independent and non-independent islands. I have witnessed the steep differences in infrastructure, service delivery, and governmental care to the citizens between dependent and non-dependent islands. A combination of national sentiments compounded by desperate impoverishment, destitution, and colonial neglect triggered independence to several islands in the Caribbean.[21] The infrastructure's quality, such as road, water, and electricity supply, and government buildings were decaying and outdated in the independent islands. The service quality and delivery of governments to their residents were poor, and I heard many voices complaining about education and healthcare. Poverty and unemployment hovered around 20% and were much higher if not for the safety valve of Caribbean emigration to Britain, Canada, and the United States. Governments were saddled with huge debts revealed in the buildings' decay and lack of providing the upkeep of the infrastructure. All aspects of the public infrastructure suffered neglect: drinking water supply, electricity provision, postal services, telephone service, and public transport.

The disastrous regime experimentation in the Caribbean, such as in Grenada, led to pessimism and distrust in democracy; and hopes for a better tomorrow were steadily eroding. A culture of lawlessness and corruption was openly debated-attributed to a loss of respect for the state and the rule of law. The delivery of state-run services failed to live up to expectations. Excluding a large percentage of the population from the economic benefits has fueled the emergence of a gang culture affecting morale, productivity, and safety.[22] Not surprisingly, the poor's experience with political disenfranchisement as a pathway to economic self-reliance and development triggered a sentiment and movement in other small islands to seek constitutional arrangements with the mother countries short

---

[21] I attended multiple meetings in the Caribbean region during the 1980s. My impression was that we were viewed as marginal members of the Caribbean region. In some instances, representatives of independent islands would be condescending with representatives or citizens of a dependent island. It was inconceivable and incomprehensible how some islands would resist sovereignty and stay with the mother country. Coming from a dependent island with resisted independence, I felt that we were treated as second class citizens in the Caribbean context because of our political status.

[22] According to The Economist, the average murder rate increased 15.7/100,000 in 2004 to 19.9/100,000 in 2007, despite relatively strong economic growth and falling unemployment in the islands belonging to the Organization of the Eastern Caribbean (Anguilla, Antigua, and Barbuda, Dominica, Grenada, Montserrat, St. Kitts and Nevis, Saint Lucia, and St. Vincent and the Grenadines). There is a deterioration in safety and security, an increase in drugs and gang crime. See: https://www.economist.com/news/2008/03/20/a-caribbean-crime-wave.

of political independence. These small islands also extracted generous concessions from the mother countries, such as subsidies to modernize their infrastructure and leading sectors of their economy and preferential access to the European market. Political affiliation seems a good predictor of success because empirical evidence suggests that these islands' political status generated a higher standard of living than in many of the other independent islands. As mentioned by Bertram (2004): "… there is no clear incentive for presently dependent island territories to seek independence, and good grounds for them to hold on [to] the status quo." (p. 353).

6.2   THE POLICY CHOICE HYPOTHESIS

While the above studies comparing the economic performance between dependent and non-dependent small islands provide empirical evidence that the dependent islands have been doing better in economic growth and prosperity terms than the independent, they cannot explain the variance across dependent small islands revealed in their empirical analysis. For example, Aruba, with a legal jurisdiction short of independence and the Cayman Islands, a British Overseas Territory in the Northern Caribbean, enjoy higher prosperity as measured by the GDP per capita compared to other politically dependent small islands such as Curacao, Bonaire, Montserrat, and the Caribbean French overseas departments (Martinique and Guadeloupe). Aruba achieved unprecedented affluence through tourism specialization (TS), while the Cayman Islands combined offshore banking with tourism.

Arguably, the difference in economic output cannot be explained by political affiliation or jurisdiction alone. Something else is likely at work to spawn this difference. The legal jurisdiction short of independence as the explanatory variable for growth, while enticing, cannot explain the existing variance in economic prosperity across dependent small islands. Other plausible explanations, such as TS and comparative advantage, may account for this variance.

My hypothesis argues that TS is the determinant factor in propelling small islands to economic and social prosperity. That tourism provokes growth in small islands is not new, and the channels through which this growth occurs are not novel either. This hypothesis was the center of attention to the study of McElroy in 2006. The study worked through seven essential data series (land area, population, stayover visitors, one-day visitors, the average length of stay, hotel rooms, and average yearly tourism expenditure) to assess a sample of 36 small islands. Based on these data series, McElroy ranked the islands and grouped them into three development levels: most, intermediate, and least. The most developed group consisted of nine islands (UK Virgin Islands, St. Maarten, Aruba, Cayman Islands, Turks/Caicos, Bermuda, Malta, Guam, and US Virgin Islands) defined by the prominent presence of tourism development in their economy and political affiliation. Eight of the nine islands in this group were politically dependent.[23]

I argue that what makes the difference is the policy choice that islands make, based on their comparative advantage. Aruba is a case in point. The island perspective in the 1980s reflected a clear understanding of the rapidly changing conditions imposed by globalization and the need for quick and flexible responses to the changes. Aruba refused to accept that growth performance depended on size and instead embraced that growth performance resulted from choice-from the policy. The rapid policy deployments to embrace tourism as the strategic shift to rearrange its international trade regime away from the once-lucrative oil industry and, in conjunction with dynamic societal adjustment, put the island back on the prosperity track.

In 1986, two years after the close of the oil refinery, Aruba's unemployment rate soared to nearly 25%.[24] The byproduct of this rate grossly impacted Arubans' quality of life. Responding to the shock required expedited strategic planning and a creative agency for Aruba to convert itself from oil to another industry that could be economically feasible, while also bearing sustainable auspices. Tourism was selected as the likely economic activity to replace the oil refinery business. The burgeoning tourism industry spread quickly to all economic and social realms spurring new businesses and jobs, and prompting immigration to the island to seize the opportunities generated by tourism development. Aruba's population grew from 56,000 to almost 100,000 from 1986–2010. Its real GDP per capita rose by nearly $18,000 between 1986 and 2010 ($6,662 to $24,595) or more than three times. By 2019, Aruba's GDP per capita was $31,633, nearly five times the 1986 level.[25]

---

[23] See McElroy (2006).
[24] The unemployment level was conspicuous, similar to the US Great Depression's unemployment figure in the 1920s.
[25] By 2018, Aruba's economy surpassed the economy of Curacao.

By appealing to and capturing foreign markets and its high-income elasticity product, Aruba was able to account for the restrictions and limitations that so many studies assigned to small island prosperity. However, booming small island tourism requires more than a foreign market and a product to sustain itself. Sun, Sand, and Sea was the initial draw for tourism. This continues to draw new and repeat visitors (a 60% repeat attendance rate has indicated a loyal following) as do the amenities of clean water, safe food, tourist safety, inviting interaction of locals to visitors, ease of currency use (such as US dollars' acceptance on the island), enticing events and activities, and transportation. The benefits of tourism have filtered into the well-being of the residents and the tourism industry's economics. Three studies that I was involved with from 2011 to 2017 corroborate the previous claim. These studies found evidence of a significant positive relationship between TS and wellbeing among the Aruba population.[26] Nearly three out of four residents indicated that they were happy and satisfied with their lives. For example, the study conducted in 2016 found that tourism had a positive impact on life satisfaction ($\beta=0.09$; p $p<0.000$) via higher incomes and opportunities.

With higher incomes, Aruba has been able to maintain the quality of its hotels and the services offered, including the infrastructure required for its patrons' continued satisfaction. Albeit, Aruba's great concern has been the impact that tourism numbers could cost the environment-the natural resource upon which its tourism depends. As a result, Aruba has taken decisive and constructive action to protect its land by assigning 20% to conservation and to foster sustainable renewable energy making the island energy less dependent on imported fossil fuels. Due to its aggressive renewable energy, Aruba was named in 2015 the National Geographic World Legacy Award Winner for "Destination Leadership" at the ITB Berlin, the world's largest tourism convention.[27] Currently, it is building and structuring a wind energy farm toward the continued efforts to account for the losses incurred by its failed oil industry as well as generating a more environmentally and economically friendly development of a wind energy farm. Aruba's economic performance does not satisfy the expectation that

---

[26] I conducted together with colleagues from the University of Central Florida. These studies collected more than 2,200 surveys in 2011, 2013, and 2016. See, Croes, Rivera, and Semrad (2011); Ridderstaat, Croes, and Nijkamp (2015); and Croes, Rivera, Semrad, and Khalilzadeh (2017).

[27] Aruba increased its energy productivity generating utility rates that are approximately $0.28 per kilowatt-hour (kWh), below the Caribbean regional average of $0.33/kWh. Please visit the link https://www.nrel.gov/docs/fy15osti/62709.pdf.

a small island is nearly consigned to economic weakness, vulnerability, and marginality. Successful economic shifts from one industry to the next to counter economic woes are not unique to Aruba. St. Lucia switched from its banana industry to tourism in the 1990s relatively smoothly.

However, what is unique to Aruba is the successful structural switch from higher productivity leading economic sector (oil industry) to a lower productivity sector (tourism) without impairing economic and social prosperity. Au contraire! The truth is that tourism propelled the small island to even higher prosperity levels. For example, Aruba's human development index (HDI) was 0.857 in 2010,[28] According to the United Nations Development Program (UNDP), Aruba is one of the highest among small island destinations and among the very high development countries in the world. By 2017, Aruba's HDI increased to 0.874.[29] The island also enacted a universal health care program providing health care to all residents; moreover, by 2010 social programs accounted for nearly 25% of the GDP, a figure commensurate with those of the OECD countries.[30] Aruba has become more munificent as it got wealthier, a phenomenon resembling Wagner's Law.[31] The Aruba case suggests that small size does not constrain a small island in the scale of its social expenditure, as anticipated by the economic literature.[32]

6.3   A STRUCTURED FOCUSED COMPARISON APPROACH

Following our argument in Chapter 2 that the impact of specialization is a choice, I applied the Dutch Caribbean to assess the policy choice hypothesis

---

[28] Aruba' HDI compares favorably with other Caribbean islands such as the Bahamas (0.789), Barbados (0.776), Antigua and Barbuda (0.774); and St. Lucia (0.714).

[29] The HDI scores are the author's estimations. By 2017, Aruba's HDI was nearly that of the very high human development value (0.890), was higher than the small island development states (SIDS, 0.722), and the average in Latin America and the Caribbean region (0.758), including the Bahamas (0.804) and Barbados (0.813). See: http://hdr.undp.org/en/data.

[30] Social programs include social spending, health, and education. See, CBS (2012). *General Government Sector of Aruba, 2000–2010*. Oranjestad: National Accounts Department of CBS. Aruba's nearly 25% of social programs puts the island at a higher level than the median of 21% of the OECD countries in social spending in 2016. Please visit: http://www.oecd.org/els/soc/OECD2016-Social-Expenditure-Update. pdf, 2019. In contrast, countries such as Chile and Mexico spent less than 15% of GDP on public social programs.

[31] See Wilkinson (2016). Wagner's Law reveals a correlation between the rise in GDP per capita and government spending.

[32] See, for example, Briguglio (1995); and Alesina and Spolaore (2003).

argument. The analysis considered several development aspects, such as initial conditions, business environment, institutions, government system, and rule of law. The Dutch Caribbean was selected as a case study to test the constitutional economics hypothesis using a structured, focused comparative case study approach.[33] The comparative case method consists of two conditions: cases should entertain similar conditions (e.g., a political dependence in the Dutch Kingdom) but should manifest different characteristics in other aspects (different stage in TS). Aruba and Curacao reveal these conditions. Both islands are dependencies in the Dutch Kingdom while revealing different levels of TS. The sample time period is 1986 to 2018. The year 1986 marked the beginning of the new political status of Aruba (Status Aparte), while 2010 marked the demise of the Netherlands Antilles, with Curacao as the political and economic capital. On October 10, 2010, Curacao acquired the same political status as Aruba in the Dutch Kingdom.

We argue that the level of TS explains the variance in prosperity between the two islands. Let us start with the economic output of each island and how it plays into the population. This comparison is predicated on the suitability of GDP per capita as the choice for comparison. The GDP per capita of Curacao and Aruba reveals an interesting pattern. In 2010, the nominal GDP per capita was $20,782 and $24,595 in Curacao and Aruba, respectively, while in 1986, the GDP per capita for these countries was $7,158, and $6,662. These figures suggest that in the past 25 years, the "catching up process" (i.e., convergence) undertaken by Aruba shows faster growth numbers than Curacao. Aruba's GDP per capita increased by 6% annually and rose from US$ 6,662 in 1986 to US $24,595 in 2010. Aruba's GDP per capita in 2019 was US$ 31,633 compared to Curacao's GDP per capita of US$ 19,216. The difference in GDP per capita was more than US 10,000, indicating a widening gap between the two countries in terms standard of living.[34] By 2019, Aruba's economic output was US$ 3,363 billion compared to US$ 3,140 billion in Curacao, exceeding Curacao's economy for the first time in history (Figure 6.1).

Table 6.1 provides insight into critical differences of economic performance in a comparison of Aruba with Curacao. The comparison between Aruba and Curacao highlights a contradiction of the alleged

---

[33] For a discussion on the predicaments of the structured, focused comparison case study approach, see George (2019).

[34] UNCTADSTAT at: https://unctadstat.unctad.org/CountryProfile/GeneralProfile/en-GB/531/index.html.

advantages granted by the political dependence and the 'economics of constitutionalism' proposition. Political dependence may propel an advantage when compared to other independent small islands. However, relying on a mother country may not warrant an economic advantage over similar constitutional configurations. With Aruba and Curacao (two countries with the same political status) Aruba outperforms Curacao within the same kingdom. This difference may be because other channels are at work. Aruba's specialization rose 5.2% a year, whereas Curacao's was 3.2% between 1986 and 2010 (nearly 25 years). This resulted in a 2010 increase in 72% specialization for Aruba and only 21% for Curacao. The economic catch-up process and outcome happened in just a one generation timespan.

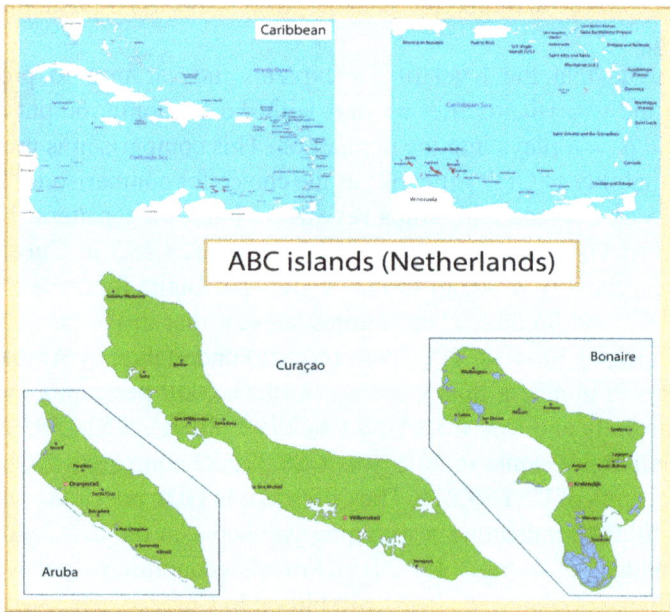

FIGURE 6.1   Map of Aruba and the other islands in the Dutch Caribbean.

Further, Aruba's per capita GDP in 1986 was US$ 6,662, while Curacao's was US$ 7,158. By 2010, Curacao's US$ 20,782 lagged behind Aruba's US$ 24,595. Around 15 years into *Status Aparte*, Aruba's growth was indicated as 8.4% and Curacao's 2.7%. Aruba grew three times faster than Curacao. At this rate, the average growth of Aruba's standard of living would double in just nine years,[35] but Curacao would require 26 years to do the same.[36] If we take the average growth from 1986 to 2010, it will take Aruba 12 years[37] to double the standard of living, while it would take Curacao 44 years,[38] almost two generations to accomplish the same thing.[39]

TABLE 6.1   Aruba and Curacao: Selected Indicators, 2010

Parameters	Aruba	Curacao
Nominal GDP per capita US $	24,595	20,782
Real per capita GDP % (1986–2010)	6	1.6
Population (× 1000)	102	142
Average population growth (1986–2010) (%)	2.26	–0.3
Area sq. km	0.19	0.96
Openness (%)	138.6	48.1
Tourism receipts (millions of US$)	1,264	380.72
Tourism specialization (%)	72	13
Average specialization intensity (1986–2010) (%)	5.2	3.2

Source: Author's own calculation from data of IMF, World Bank and Central Banks. Openness refers to an economy openness to the outside world and is defined as the ratio of the sum of exports and imports to the GDP of a country.

Table 6.1 also reveals the difference between Aruba and Curacao in their respective degree of openness. While Curacao displays 48% openness, Aruba indicates an openness of nearly 139% (a ratio of nearly three to one). Aruba's real GDP per capita growth was nearly four times faster than Curacao. Aruba's higher degree of openness may suggest the reason this island had more robust economic growth than Curacao. From an empirical perspective, the connectedness of trade openness with economic growth indicates that countries that propagate outward-oriented policies display healthier economic evolution. For example, Seetanah found in a panel of 19 island economies from 1990 to 2007 that openness was an essential

---

[35] $(1.084)^x = 2 \rightarrow \log(2)/\log(1.084) = 8.59$ years to be specific.

[36] $(1.027)^x = 2 \rightarrow \log(2)/\log(1.027) = 26.017$ years to be specific.

[37] $\log(2)/\log(1.06) = 11.89$ years.

[38] $\log(2)/\log(1.016) = 43.67$ years.

[39] Pinker (2018) claims that since 1995 only 30 countries out of the 109 developing countries have achieved economic growth rates that propelled a doubling of their income within 18 years. Aruba doubled its income in just nine years.

element of growth, and estimated a growth elasticity of 0.21. Seetanah's study also revealed that tourism in small islands has higher growth effects compared to other countries.

This impressive growth of Aruba occurred while simultaneously experiencing an annual population growth of 2.3%. Aruba's population almost doubled in the past 25 years. Curacao grew at a much lower pace depicting an annual growth rate of 2.1%. However, at the same time, the island experienced negative population growth during the period under consideration (–0.3%). Much of the Curacao population migrated to the Netherlands, dropping from 151,000 at its peak in 1986 to 128,000 in 2004.[40] The contrasting population experience in Aruba and Curacao corresponds to McElroy and Parry's findings, referencing the intense emigration to the most tourism penetrated islands in search of jobs and opportunities.[41]

Although this was the case for Aruba as a receiving country, Curacao people did not opt to migrate to the most tourism penetrated island but instead moved to the Netherlands. So, Curacao was an exception, perhaps because of the existence of a mother country. This result appears to contradict the proposition of advantage through the legal jurisdiction mechanism. The variance in economic performance between Aruba and Curacao-two dependent islands within the same Kingdom, and yet with economic output differentials-cannot be explained by this variable, i.e., the political dependence status.

6.4   THE VOLATILITY OF SPECIALIZATION

However, in examining the opposition to TS, the large difference in the degree of openness would suggest that Aruba would be more susceptible to external shocks than Curacao. The downside of trade openness is economic volatility spurred by external shocks. These shocks may, in the end, hurt the island economically offsetting all its economic gains, such

---

[40] In economic downturns, it is typical that residents would migrate to the Netherlands, searching for better opportunities and taking advantage of the special relationship with the Netherlands. This economic and social escape valve (outflows of workers) serves as an equilibration of per capita income levels.
[41] See McElroy and Parry (2010). Migration has been a constant phenomenon for small islands' residents moving to metropolitan countries searching for opportunities and jobs. Migration has functioned as a social safety valve when opportunities are scarce on the islands. Those leaving the islands returned or have sent remittances to their island, creating an important economic pipeline to families who stayed behind, thereby prompting economic activities and opportunities on the islands.

as increased productivity and efficient allocation of resources all of which were likely triggered by trade openness.[42] Volatility may incite a welfare loss through the negative effect of uncertainty. In other words, trade openness could have beneficial as well as detrimental effects. For this reason, some consider tourism as a "fickle" industry. Indeed, when we estimated the coefficient of variation (CV) for Aruba and Curacao's real GDP, the results indicated that Aruba, from 1986 to 2010, had higher volatility than Curacao. Aruba's CV was 2.1644, while Curacao's CV was 1.51622 (both in logarithm).

It seems that islands with a higher dependence on tourism may experience higher volatility and, therefore, tourism could hamper growth according to trade theory. It is critical at this juncture to remind the reader about the main tenets of trade theory before engaging in the empirical assessment of the impact of volatility on growth.[43] Trade theory suggests that considerations for comparative advantage are the basis upon which nations move exports. The expectation is that the comparative advantage will bolster the nation's economics. This would indicate that the degree of openness via exports would establish confidence in a country's economic productivity.

Insofar as small countries can experience a small domestic market, it theoretically becomes critical to secure a sustainable position in the international trade market. This positioning nurtures the domestic market while extending opportunities to gain the necessary financing required to procure the types of imports essential to the country's welfare and its determined development. Further, international trade and the competition it imposes foster greater efficiency of productivity. Thus, trade theory asserts that there exist substantial and positive impacts that exports encourage: for example, economies of scale, ensuring positive externalities in areas not directly involved in export, the purposeful and proficient apportionment of resources instigated by the intensification of competition, a surge toward effective and efficient research and development, the recognition of investing in human capital, and, finally, the more favorable policies regarding foreign exchange constraints.

You might now ask what the connection is between the discussion of trade, tourism, and growth. Trade literature supports the idea that openness

---

[42] The assertion is that tourists specializing islands would be prone to external shocks, and hence tourism could cause economic harm to the island. See, for example, Adamou and Clerides (2009).
[43] In Chapter 7, I will empirically assess the volatility induced by openness and tourism on growth.

can surmount the limitations that small domestic markets pose in that capturing international trade can supplant scale constraints; likewise, the level of international competition filters to a more efficient productivity process.[44] Quantifying trade openness is accomplished by summing both nominal imports and exports and then dividing by the nominal GDP. For small countries, the degree of openness, as indicated by their trade shares in GDP exceeds 100%. However, the degree of openness does not account for its distinctions in income growth for the Caribbean.[45] As a result, it is possible that how germane a country's trade openness is becoming relative to the types of exports rather than the quantity of its exports. Other studies also indicated that the degree of openness does not serve to account for the Caribbean's income growth.[46]

Notably, where it appeared that Aruba could be more susceptible to external shocks than Curacao, it seems Aruba has overcome the penalties that stem from volatility. The island appeared (annual average real GDP per capita growth of 6% over 25 years) to have weathered the higher volatility without affecting its long-term growth trajectory (see Figure 6.2). Aruba's very high HDI of 0.857 attests that despite its economic vulnerability due to its TS strategy, Aruba has enjoyed a high income per capita and a very high human development. Its HDI score is higher than that of Curacao at 0.800.[47]

There are two plausible explanations for the differences in economic output and patterns between these two islands. The first plausible explanation is that tourism inflows as measured by tourist receipts have revealed relatively low volatility. For example, Aruba's CV for tourist receipts was 12.89 compared to 62.84 for Curacao. Aruba's CV is also one of the smallest coefficients of variation in the Caribbean, reflecting a relatively stable tourism revenues flow. Table 6.2 depicts a comparison among a selected group of Caribbean countries, where Aruba shows the second smallest coefficient after St. Vincent. This low level of Aruba's tourist receipts CV may indicate the strong destination appeal Aruba enjoyed from the tourist markets due to its high-quality product.

---

[44] See, for example, Krueger (1985).
[45] See Croes (2011). The small island paradox.
[46] See, for example, Wint (2002).
[47] See CBS (2014). *Human Development Index* (Vol. 1, No. 2, pp. 28–30). Modus. http://digitallibrary.cbs. cw/content/CB/S0/00/01/55/00002/Modus%20jrg%2012%20nr%201-2.pdf. A study from Gatt (2004) suggests that HDI tends to be relatively high for very small countries.

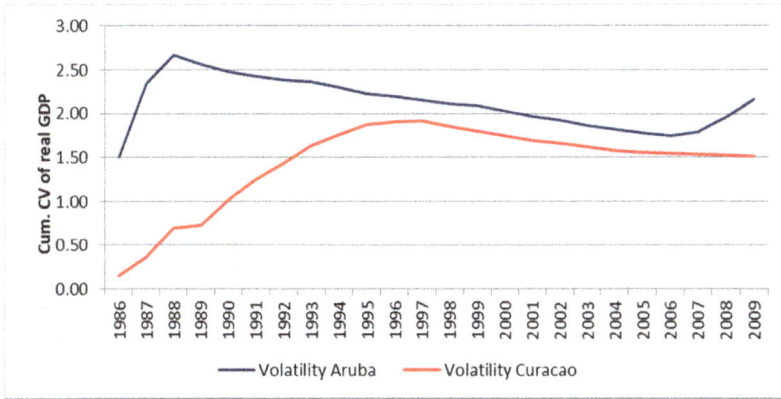

FIGURE 6.2   Cumulative volatility real GDP Aruba and Curacao.

TABLE 6.2   Average Growth Rate and Coefficient of Variation of Selective Caribbean Destinations, 1995–2007

Country	Average Growth Rate (in %)	Coefficient of Variation
Aruba	0.02030221	1.0497
Bahamas	−0.00332960	1.4498
Barbados	0.02298961	1.3859
Dominica	0.02679489	1.9140
Grenada	0.02321313	3.6772
St. Kitts	0.04353641	2.2506
St. Lucia	0.01955889	4.5839
St. Vincent	0.03613008	0.8676

Source: Adapted from: Croes (2011). The small island paradox.

The second plausible explanation is the Aruba economy's higher productivity performance compared to the one in Curacao. Aruba's average TFP from 2010 to 2017 was **98.9** compared to 97.1 in Curacao.[48] According to the International Monetary Funds, five of the seven years

---

[48] The Total Factor Productivity (TFP) is what is left over when we accounted everything in the economy that can be measured. The estimation of the TFP growth used standard Cobb-Douglas production function with an Index TFP 1996=1. While Aruba experienced a slight productivity growth increase during that time, its TFP growth has declined since the 1990s. The decrease in productivity growth may stem from tourism specialization as a lower productivity growth sector compared to manufacturing, the negative external shock of the Great Depression, or the decline in labor's contribution. TFP growth, while seemingly low, is very close to the USA TFP growth average of 94.0 since 2004. See, for example, Banerjee, and Duflo (2019). The slow TFP growth is ominous to the slowing economic growth in the USA and EU countries.

since Curacao gained autonomy in 2010 were recession years. From 2010 to 2017, Curacao grew at an annual average growth rate of 0.004 compared to Aruba of 1.13 during the same time span. This average growth rate occurred despite September 11, the Great Recession, and the closure again of the Aruba oil refinery in 2013.[49] Every time, the island bounced back, regained growth, and overcame volatility and vulnerability.

Aruba has been consistently ranked among the top tourist destinations in the Caribbean. A study I conducted in 2011 ranked Aruba at the top among the Caribbean islands. The ranking was based on a productivity model that included three main components: the current performance in the global tourist market scaled by size, the dynamism of performance over time (growth rate), and the size of the industrial base in the economic structure. The ranking is revealed in Table 6.3. Spending per tourist and tourism value-added seems to spur the top position of Aruba in the ranking. Note that Curacao is almost at the bottom of the ranking, which suggests that Curacao's tourism product reveals lower quality offerings.

The implications are twofold. First, functional openness rather than structural trade openness, seems to spur growth. As remarked by Armstrong and Read: "Although structural openness is a consequence of their small size, functional openness is the outcome of a conscious endogenous policy choice." Second, destinations with higher quality products seem to grow more rapidly. Paradoxically, the often-cited vulnerability induced by openness turned to be Aruba's strength because openness compelled the economic structure and activities that allowed Aruba to become internationally competitive in open markets while preventing Aruba from sliding into protectionism on economic terms, much the same as so many other islands.

6.5   THE POLICY CHOICE: SPECIATION[50]

Fostering development to substantiate growth rates is an apprehensive undertaking. What is not part of the undertaking is the geographical size

---

[49] The Valero refinery operated the Exxon refinery from 1990 till 2013, when it seized its operations. The Exxon refinery closed in 1985. See, for example, Ridderstaat (2007) for a history of the oil industry in Aruba.

[50] Bertram and Poirine (2007) coined this word to depict the intensity and degree of specialization. They opined that economic "speciation" is a conscious decision by a small island to specialize in one or two leading economic sectors.

TABLE 6.3 Spending per Tourist and Value Added Driving the Ranking in the Tourism Competitiveness Index, 1986–2007

Country	Tourism Competitiveness Index	Rank	Tourism Receipts per Capita	Rank	Growth Rate of Tourism Receipts	Rank	Tourism Value Added Ratio of GDP	Rank
Aruba	0.752	1	$10,960	3	0.087	3	0.680	2
British Virgin Islands	0.751	2	$19,864	1	0.067	9	0.441	8
Anguilla	0.639	3	$6,308	7	0.084	5	0.671	3
US Virgin Islands	0.623	4	$13,573	2	0.055	12	0.559	4
Antigua and Barbuda	0.442	5	$4,088	8	0.035	16	0.789	1
Cayman Islands	0.428	6	$7,060	4	0.073	7	0.253	13
Bahamas	0.395	7	$6,466	6	0.038	15	0.531	5
Guadeloupe	0.394	8	$553	17	0.091	1	0.317	11
Saint Lucia	0.392	9	$2,225	10	0.059	11	0.529	6
Grenada	0.357	10	$645	16	0.086	4	0.276	12
Dominica	0.329	11	$800	14	0.082	6	0.243	14
Saint Kitts and Nevis	0.318	12	$2,140	11	0.053	13	0.421	9
Barbados	0.316	13	$2,638	9	0.046	14	0.457	7
St. Vincent Vincent and Grenadines	0.309	14	$842	13	0.065	10	0.350	10
Martinique	0.298	15	$675	15	0.09	2	0.103	16
Curacao	0.241	16	$2,037	12	0.068	8	0.03	17
Curacao	0.241	16	$2,037	12	0.068	8	0.030	17
Bermuda	0.136	17	$6,550	5	0.017	17	0.130	15

Source: Adapted from: Croes (2011). The small island paradox.

of a destination or its jurisdiction's independence. We already alluded to Wilkinson's 1987 deterministic view, which states that their size constrains small island countries. Therefore, tourism is the only "inevitable" option under which the country can legitimately strive for development and growth. However, if this is so-if small countries are relegated to one logical choice, how do we account for some small islands' success over that of others? The question is: if size or independence cannot account for growth variance in small island destinations, what can?

Perhaps part of the answer is identifying how small island destinations determine their path to growth and development? Chapter 2 made the point that specialization is a matter of choice. Small islands may choose to venture into the global market through openness, showing no fear, but may do so imprudently without confronting vulnerability and volatility or not. They could choose to specialize in tourism or not. If choice matters, then the difference in economic performance between independent and non-independent small islands should be associated with specialization choice. As indicated before, the specialization choice is not so much the choice to specialize, but what to specialize in and the degree of specialization.

According to Clancy's study in 1999, there exist factors influencing small island options that must be considered, understood, and respected. He contends that the range of ideas and preferences, as well as orientation, must characteristically be determined at the time an option is considered and selected. Another part of the answer, according to Clancy, lies in small island government as to whether or not it contributes to the capacity of growth and development via the choice made. He also contends that the success or failure of choice is a consequence of the selected industry itself. Thus, growth, and development are impacted by the sum of the parts of the industry. This view corresponds to the emerging notion that small islands should strategically select their economic specialization niche to compensate effectively for their small size via optimal endogenous policy formulation and implementation.[51]

In a study in which I collaborated with Ridderstaat and Nijkamp, we found that Aruba became more dependent on tourism beginning in 1986. The upward and downward fluctuations in the GDP output were quickly adjusted towards equilibrium compared to when the oil industry

---

[51] See Armstrong and Read (1988, p. 13).

played a significant economic role. This means that Aruba could adjust to economic shocks faster as it got more specialized in tourism. The study also found that the tourism and GDP output was cointegrated and that a 1% growth in tourism receipts would spur a long-run GDP growth of 0.82%, ceteris paribus. TS explained 99% of the variance in economic growth (F = 17.1100; p = 0.0000).

To compare with Aruba, I formally tested the association between TS and economic growth regarding Curacao. The quantitative analysis is based on a simple econometric framework relating to the impact of specialization on the level of real per capita GDP growth. TS is defined as the ratio of stayover visitors (tourists) to the local population.[52] The log of the annual real GDP growth rate (i.e., the percentage change) was regressed against TS. Log was not taken for the TS index, so the time series underwent a semi-log transformation. The time series data were integrated in the first order, and TS and real GDP growth were cointegrated in the order of one (I(1)), guaranteeing the significance of the statistic test usually conducted in OLS regressions.[53] TS explains 74.67% (F = 9.8415; p = 0.0001) in the variance in real GDP growth with Curacao. However, tourism only had a positive short run significant effect at 10%, while it did not have any significant long run effects on the economy.

So, what explains Aruba's steep growth compared to Curacao? The difference in economic performance between the countries can be examined empirically by analyzing the role of specialization (policy) and considering the impact of this factor on real income growth and volatility. We already mentioned the difference in economic output between Aruba and Curacao. Along with this difference is the high volatility that Curacao experienced regarding tourism receipts 1986–2010 compared to Aruba. To measure the volatility, we calculated the coefficient of variation (CV), which is the ratio of the standard deviation to the mean. The CV with Aruba was 12.89, while Curacao was 62.84. Higher volatility is

---

[52] The literature has defined multiple indicators of tourism specialization, measuring a particular feature of tourism. However, there is no consensus regarding which of these indicators is the best to use. This definition varies somewhat from my initial definition as the ratio of tourist receipts to the GDP because of calculation problems such as autocorrelation and heteroscedasticity. The small sample size could also have played a role when applying an ARDL model.

[53] Note: Critical values are based on Narayan (2004), with unrestricted intercept and no trend, for the number of regressors = 3 and number of data points. The significance was at the 1% level. The study applied an ARDL regression model to assess the relationship between tourism specialization and real GDP growth.

associated with lower economic growth, while lower volatility is associated with higher economic growth. Aruba's lower tourism receipts
volatility seems to compensate for the slightly higher GDP volatility
compared to Curacao.

Why is Aruba choosing TS, while Curacao persisted on import substitution, prolonged trade barriers, and diversification? In my conversations
with multiple officials and residents in Curacao, I noticed a clear appre-
hension regarding tourism. The belief was that tourism was not a real
economic industry but an economic activity not worth the investment and
engagement of the local population. Arguably, Curacao lacked the will to
forge a tourism pathway for future growth. Serving others did not seem
high on their DNA list. Specializing in tourism seems inconsistent with
the long-held belief in Curacao that diversification was a better development strategy for prosperity because it was believed that diversification
could fend off the island from economic shocks making it less vulnerable. However, diversification strategy may distract a small island from
focusing on flexible deployment of its resources because openness has
its toll and costs: volatility. Curacao experienced steep volatility in its
economic growth from 1986–2010.

Diversification may take the eye from the economic growth pole, as
suggested by a study of Bertram and Poirine.[54] According to their study,
diversification erodes the focus and rapid response capabilities required
to face the ebb and flow of the external environment (due to openness),
thus depriving the island of taking swift advantage of external opportunities. This lack of flexibility and the rapid response seems to be what is
taking place in Curacao, where the four economic pillars of the island, i.e.,
offshore banking, ship repair, oil refinery, and tourism, have experienced
steady erosion.

Openness to the international market is relevant as a development
strategy. However, while exporting, for example, may not provide enough
impetus for growth, the export product could be growth-enhancing. This
follows Demas's proposition that structural openness resulting from scale
problems differs from the functional openness, which results from a
deliberate policy choice (Demas, 1965). Only tourism has shown positive
performance, albeit not strong enough to prompt the economy out of its

---

[54] See Bertram and Poirine (2007). Their study advocated for small islands to embrace specialization as
a development strategy, calling it "speciation," a deliberate choice towards development.

lull. However, the latter has not received the necessary and swift support and attention to sustain the economy of the island.[55]

Aruba chose the path of TS in 1986. The failure of the oil industry in the 1980s forced the government to ascertain its direction for economic recovery and to do it quickly. Selecting tourism as a means of economic recovery was bold in that the 1985 IMF report indicated that tourism's potential to restructure the economic crisis into a profitable and sustainable enterprise was negative. Indeed, a 50% supply-side increase of hotel rooms would not suffice toward the 1980s disaster's amelioration. IMF challenges included concerns regarding sharp losses in government revenues and disposable foreign exchange earnings, which would drop 42% below those of 1982. Nonetheless, Aruba persevered, met, and conquered the challenges, achieving astonishing success in less than 25 years.[56]

The choice for tourism development was the only strategic option after the closing of the refinery. Aruba had the infrastructure in place, had the beaches, had the human capital to accelerate tourism development, and had the will to experiment and allocate the resources to create jobs, profits, and growth. There was a clear consensus that tourism should become the new engine of growth, and the government and the private sector embarked on designing and implementing programs about foreign direct investment, imports of technology-laden goods, export-driven competitiveness and magnitude, the challenges, and shortcomings in maintaining competitiveness and sustainability, and ameliorating market failures. Tourism attracted the best resources and talents, and capital allocation was yielded toward the tourism sector for its continued improvement.

The promise of tourism also generated optimism for potential prosperity. Such was Aruban pride in gaining the new *Status Aparte* that Aruban residents, *en masse,* exchanged their Antillean guilders to the newly minted Aruban florin. They enacted this in the first three months of 1986 to introduce the Aruban florin a financial success. This remarkable achievement happened despite all the surrounding uncertainties such as

---

[55] In 2010, Rosen College Dick Pope Sr. Institute *for* Tourism Studies alerted the Tourism Commissioner of the lack of a clear vision for tourism on Curacao's economic development strategy. See, Croes, Rivera, and Tesone (2010). Report to the Commissioner of Tourism of Curacao. Dick Pope Sr. Institute *for* Tourism Studies: Orlando, FL. This lack of clear vision bumps into the prominent tourism role that Curacao played in the 1950s and 1960s. Because of this prominent role, Curacao may have been chosen as the venue for creating the Caribbean Tourist Association in 1951, the precursor to the Caribbean Tourism Organization.

[56] See, Croes (2010).

the new political status, the closing of the oil refinery, the shrinking of the economy by almost 40%, and the initial boycott of Willemstad and The Hague.[57] Aruba's progress and prosperity were not the results of magic, but problem-solving and its ability and skills to practice collective action and labor productivity shifts. In short, Aruba was able to eschew the common automatic bias in favor of the status quo and shook off the persistence of habits favoring the future instead of the presence.

On the other hand, Curacao may have been marred in the status quo and chose to muddle through as a strategy despite the looming dangers for its oil industry and financial center.[58] The island was not forced to change course, and policy-makers failed to examine and conduct experimental options regarding Curacao's future. Muddling through may have been the result of the persistent political instability after 1969. A study comparing the political stability between Aruba and Curacao using the Pederson Index of political stability found that the years spanning 1979 to 1994 revealed an index of 25 for Curacao and 10 for Aruba.[59] Aruba's stability index was commensurate with those of European countries, which hovered around 9.

This political instability continued after 10-10-10 when Curacao had four Prime Ministers in just three years. For this island, the lack of experimentation and the degree of experimentation has been turning out to be costly from an economic and social cost. Policy habits and the focus on the present instead of the future shaped Curacao's opportunities with seemingly dire consequences. Maintaining a status quo by retaining a policy was to forward and maintain an unstable system fraught with beliefs and confidences in existing premises that held no promise for the

---

[57] See, for example, Croes and Moenir-Alam (1990). Hirschman (1984) refers to pride and love as essential resources which he characterizes as the ability of a collectivity to practice benevolence, love, and the public spirit to support and enhance the social objective and good. Institutional demands on civic behavior were instituted for the social good as long as they were not excessive and could be shown to lead to the greater good. Aruba's increased attention to community values and benevolence (e.g., pride and trust) rather than self-interest seems to have played a role in the rapid deployment of resources and talents to spur tourism specialization and prosperity.

[58] That Curacao had impending and severe issues in its social and economic areas was clear by 10-10-10. They included impending debt repayments, depressed economic growth, lingering oil refinery issues, legislation over offshore financing that destabilized the financial sector, and 15% unemployment rates with 35% unemployment for its younger generation. See, for example, the Ministry of General Affairs of Curacao (2013).

[59] See Cijntje, G., (1999). *Electorale Instabiliteit op Curacao.* https://pure.uva.nl/ws/files/1513913/ 108579_UBA003000138_005.pdf.

future. There is truth in the belief that if nothing changes, nothing changes. This likely escaped Curacao.[60]

While TS can explain Aruba's growth, it cannot explain the channels that prompted this growth.[61] Moreover, not all small islands would have the resources, the will, or the commitment to deploy quickly to take advantage of the opportunities that tourism can afford. Beyond this, there is more to determine specialization's characteristics and their importance for continued economic growth. Whether it can continue to provide high returns or reduce diminishing returns for small island destinations are essential issues to discern. Is tourism capable of performance levels sufficient to sustain small island economics through the 21st century? Chapter 7 will consider the answers to this question.

KEYWORDS

- **African Caribbean pacific**
- **Caribbean basin initiative**
- **constitutional economics**
- **European Union**
- **human development index**
- **United Nations development program**

---

[60] I am not advocating for specialization only. Instead, I am suggesting that specializing in tourism can be a vehicle that can quickly restore and propel economic activities, improving the standard of living of small island residents as the case of Aruba suggests. Arguably, tourism can become a platform of economic diversification because it touches through its externality prowess a range of other economic sectors. See, for example, Croes' Small Island Paradox (2011); and Lejarraga and Walkenhorst (2007).
[61] Naturally, Aruba has its problems as any other country. However, there is no doubt that tourism specialization triggered unprecedented growth and human development in the past 30 years.

CHAPTER 7

MAINTAINING THE GROWTH MOMENTUM

Empirical evidence indicates that tourism specialization (TS) stokes economic growth, gainful employment, and upward social mobility in small islands. The small island's fight for survival through TS brought small islands a new sense of hope and unity in their struggle and success to overcome scale. In the process, research evidence emerged regarding the relationship between economic growth and small island inhabitants' well-being.[1] That said, the question is, for how long? In other words, can TS serve as a catalyst and foundation for sustained prosperity? Sustained prosperity is an important issue because the lack of growth means stagnation, a deterioration of the population's standard of living, and lost unity. An alarming turn-around in the growth reprieve of these islands might bring instability, suffering, and pain. A declining economic pie and material foundation of small islands, or, for that matter, any country, means that there is less for each one to share and enjoy. A loss of economic opportunities may lead, as it usually does, to social dysfunction, to broken families and despair, and ultimately to an erosion of the social fabric and cohesion of small islands.

7.1 THE RELEVANCE OF ECONOMIC GROWTH

Since the time of Enlightenment, it became clear that growth is created. It does not fall from heaven, neither is it gifted nor inherited. Growth results from innovation, resourcefulness, labor, and working cooperatively toward a specified goal. Networks of people come together and configure useful

---

[1] See, for example, Lanza (1998); Lanza and Pigliaru (2000); Durbarry (2004); Narayan (2005); Neves-Sequeira and Campos (2005); Brau, Lanza, and Pigliaru (2007); Holzner (2011); Croes (2011); Seetanah (2011); Ridderstaat, Croes, and Nijkamp (2014); and Croes, Ridderstaat, and Van Niekerk (2018).

means and products that might better serve to develop and/or improve the industries that could then nourish resident welfare. Indeed, the creation of growth bodes well for the economic welfare of the nation as a whole. Such growth is crucial in that it offers citizens the agency to expand their range of human choice from which they acquire the opportunity and means to pursue a life path that accords with their values. However, producing growth and wealth is hard. Several things need to work well for successful growth production. It is like the premonition of Leo Tolstoy in his book *Anna Karenina* when he wrote: "Happy families are all alike; every unhappy family is unhappy in its way." Tolstoy alleges that for a family to be happy, vital aspects must be in place or given including good health of all family members, acceptable financial security, and mutual affection for a family. Any aspect that is missing may derail the family's operation and ultimate happiness. Arguably, and likewise, growth, and wealth acquisition should include all required factors to happen.

Take, for example, Adam Smith's specialization idea as the source of productivity. Smith indicated that through the example of the pin factory, where each worker centers his efforts on one task only, more pins could be produced than when a worker engages in all tasks to produce pins-sort of a jack-of-all-trades and master of none concept. This division of labor seems at the heart of growth and wealth. However, there is more to growth production than just a division of labor. Producers should know that consumers might be interested in purchasing the product and learning how to get the product to those consumers. The interested consumers should be convinced of the product's quality and assured that it would be delivered. Thus, a middle person should be in place to vouch for product quality and delivery. A consortium of trust must exist between all essential parties, as infringements (rule-breaking, cheating) on that trust would be detrimental to the growth process. Without a distribution person to whom one can sell the product or without trust in the system, there may be little incentive to specialize in the first place.

Coordination, trust, and scale must happen almost simultaneously for specialization to have an opportunity to grow and expand personal choice. These key aspects are common to all countries. However, small islands have a substantial additional concern, which is scale. That is why (in the case of small islands) producing growth and wealth is incredibly hard. I already alluded to this reality in Chapter 4. Thus, a top economic question results: in what does the island specialize? These economical, organizational problems stem from the need to incentivize (motivate) and coordinate human activity provoking productive activities. The problem is to decipher the nature of the most optimal societal arrangements and understand the determinants. These arrangements occur on a small island in a constrained riddled environment (including small scale, limited resources, land availability, incentives), and coordination constraints.[2] These constraints shape what choices are available, what to produce, how to produce, and for whom to produce. These three items should be resolved within an ecosystem that is perforated with the kinds of constraints that could pierce the viability of successful specialization.

After World War II, and in particular, since the 1970s, economic growth received heightened attention. Lewis, Kuznets, and Chenery considered the search for sustained economic growth vital for the advancement of the developing world. The productivity slowdown in the developed world in the 1970s, the lost decade due to the lackluster performance in the developing world in the 1980s, and the collapse of the communist world in the 1990s triggered renewed interest in the sources of economic growth.[3] The research found that resource allocation is a critical source of economic growth. But resource allocation is a hard and costly task requiring choice, adjustment, adoption of new technologies and processes, assimilation of production inputs by firms, learning competition, and even job loss. Human capital, quality of government institutions, and education appeared to be critical in prompting economic growth.[4]

7.2   TLGH AS AN ECONOMIC GROWTH SOURCE

Throughout the previous chapters, I argued that tourism specialization (TS) could provide a growth-creating capacity. Indeed, tourism literature claims the existence of a positive relationship between tourism and economic growth. There is growing evidence supporting this claim on theoretical grounds. This claim follows the tourism-led growth hypothesis (TLGH) underpinnings, a variant of the export-led-growth strategy (ELG). The TLGH asserts that economic growth over time is due to an increase in a country's tourism export. This hypothesis follows four basic ideas: tourism

---

[2] See, for example, Candela and Figini (2010).
[3] See, for example, Banerjee and Duflo (2019), in particular their chapter five.
[4] See, for example, Easterly (2001).

is a non-technology intensive sector that promotes long-term growth; the tourism growth capacity lies in triggering enough capital flows to ensure the capital accumulation; international competition spawns' efficiency among tourism firms, and tourism promotes increasing returns to scale.[5] Tourism-based economies experience inherent lower productivity. However, this lower productivity could be overcome by increased specialization in tourism. Increased specialization could improve the terms-of-trade, which can compensate for the loss in productivity in the long run. For example, inbound tourism means a consumption stimulus to the local economy, prompting more local production and employment. Increased consumption and production contribute to the balance of payments and sparks activities in other economic sectors.

The TLGH emerged with the work of Balaguer and Cantavella-Jorda in 2002. Their work examined the Spanish case, and they found a long-term positive relationship between tourism and economic growth. Their study suggests that TS financed Spanish economic structural change by generating the foreign exchange necessary to import essential goods, especially importing intermediate and capital goods. Balaguer and Cantavella-Jorda's study was six years later, corroborated by a study done by Hernandez-Martin, which demonstrated that tourism triggered economic growth by shifting the economy away from agriculture in the Spanish case.[6] Spain is a case in point where tourism played a pivotal role in assisting the country in its takeoff efforts in the manufacturing industry. The relevance of the manufacturing industry is its leading role in spurring and sustaining economic growth. This manufacturing perspective was already discussed when Chapter 5 perused Kaldor's view.

The Spanish case reveals that tourism propagates structural shifts across macroeconomic sectors, often replacing traditional agriculture sectors. These structural changes refer to modifying the relative importance of economic sectors triggering the economy to more advanced levels. Tourism revenues generate income, jobs, and investments throughout the economy, affecting sectors directly or indirectly linked to the tourism experience. These widespread tourism effects stimulate economic activities that ensue economic growth. In other words, tourism generates externalities through the tourism demand changing the production of a country's final output, and sparks a structural change shifting the economic activity favoring

---

[5] See, for example, Sinclair and Stabler (1997).
[6] See Balaguer and Cantavella-Jorda (2002); and Hernandez-Martin (2008).

either the physical or human capital accumulation. For example, my research over the years shows how tourism has transformed the Caribbean region's economies, as well as that of the Mediterranean.[7]

The empirical tourism literature suggests that small size is a valuable source that triggers the positive relationship between TS and economic growth. This positive relationship resonates with a long list of empirical studies.[8] For example, the study of Narayan, Narayan, and Prasad about a group of Pacific Islands-PICs-which included Fiji, Tonga, the Solomon Islands, and Papua New Guinea, found that a 1% increase in international tourism increased the GDP by 0.72% in the long-run, and 0.24% in the short-run. Holzner examined the relationship between tourism and economic growth based on a sample of 134 countries over nearly four decades (from 1970–2007).[9] This study is unique because it explored patterns of complementarities and substitutability between inputs into the production function. The study revealed four critical findings. First, the study found that tourism specialized countries achieved high average economic growth, grew faster than others and that these patterns have long-term effects. This positive relationship underpins the TLGH. Second, the study also found that tourism has direct and indirect long run effects via physical and human capital on economic growth and that physical capital and tourism capital are complementary. Third, countries with higher TS reveal higher levels of investment and secondary school enrollment. Fourth, countries with higher TS have low levels of foreign exchange distortions.

Commentators analyzing the relationship between TS and economic growth found similar results to those of Holzner. Overall, these studies found a positive relationship between them, so countries' economies grow when tourism activity in these countries increases. For example, a recent study of Cannonier and Burke exploring tourism contribution to economic output among 15 Caribbean countries from 1980 to 2015 showed that tourism has a significant impact on economic growth.[10] Another study by Seetanah found that tourism significantly contributes to the economic

---

[7] See, for example, Croes (2011); and Croes, Ridderstaat, and Van Niekerk (2018).
[8] See, for example, Durbarry (2004); Brau et al. (2007); Lanza and Pigliaru (2000); Narayan et al. (2010); Seetanah (2011); Croes (2013); and Ridderstaat et al. (2016). Sequeira and Maçãs (2008); and Chang et al. (2012) found that developing countries and destinations with a weaker economy seem to benefit significantly from tourism development. More recently, De Vita and Kyaw (2017) found that economic development is a source triggering a positive relationship between tourism specialization and economic growth.
[9] See Holzner (2011).
[10] See Cannonier and Burke (2019).

growth of island economies. However, similar to the other studies, tourism's economic growth impact seems modest: a 1% increase in tourism increases economic growth by 0.04 to 0.1%. This modest impact is also present in Sequeira and Nunes, Arzeki et al., and Fayissa et al.

A modest tourism impact on economic growth is also the result of a recent study by Roudi, Arasli, and Akadiri.[11] Their study covered a panel of 10 small islands (SIDS) from 1995 to 2014, and their finding indicates that an increase of 1% in tourism earnings triggers the average GDP by 0.16% in the short run and 0.32% in the long run. Holzner's study also revealed modest direct results of the impact of tourism on economic growth. However, his research also disclosed that significant tourism impact on economic growth is through indirect transmission channels. The sum of direct and indirect effects is more than double the direct impact. The most crucial indirect transmission channels are human capital, representing nearly half of the total (direct and indirect) tourism contribution.

The brief discussion about the short and long run impact of tourism on economic growth suggests that this relationship varies across small island economies. The uneven small island economic performance triggered by tourism seems to respond to more factors than TS.

7.3   SLUGGISH GROWTH

Alternatively, the empirical literature also reveals pushback to the TLGH as a long-term development option.[12] This pushback stems from three critical perspectives. The first one refers to the lack of any significant relationship between TS and economic growth. The second view claims that the effect of TS on economic growth is modest and transitory. Moreover, the last point of view asserts that TS has a non-linear impact on economic growth, which means that tourism may have a short-term positive effect on economic growth that would dissipate over time.

The first group of studies negates the existence of any long-term effects between TS and economic growth. The study of Figini and Vici investigated the long-term sustainability of TS. Their study, which focuses on 1990–2005, found no statistically significant relationship between TS

---

[11] See Roudi, Arasli, and Akadiri (2019).
[12] See, for example, see Chang et al. (2012); Figini and Vici (2010); and Adamou and Clerides (2010) contend that the relationship between tourism specialization and economic growth is not linear.

and economic growth. They attributed this difference compared to previous studies to periods under consideration: TS in small countries (2.26%) grew faster than other small countries (1.22%) during 1980–2005; however, when compared to the 1990–2005 period, tourism countries grew slower (at 1.88%) compared to other small countries.

The latter grew on an annual average of 2.52%. According to this study, TS has a higher impact on economic growth at a lower level than a higher level of specialization, suggesting that tourism is not beneficial over time. Another study by Palmer, Ibañez, and Gomez reached a similar conclusion because tourism cannot expand forever. Indeed, if the models only consider the increasing volume (arrivals), it is reasonable to expect that small islands will face inevitable deleterious effects due to size constraints (e.g., land availability).

The second perspective states that tourism development is only a temporary occurrence. Its development is not sustainable—growth results from the intensive and increasing use of one factor of production: natural resources. Once the employment of this factor reaches its maximum usage, labor productivity will become crucial in determining growth, and as a result, tourism countries will grow more slowly than others. From this perspective, tourism is not a viable development path in the long run. Three perspectives entertain this view. The first one argues that there are diminishing returns in the case of TS. The second one pertains to the lack of statistical significance between TS and economic growth, which means a lack of cointegration exists between the two constructs. The third group claims Dutch disease's existence in the relationship between TS and economic growth.

The Adamou and Clerides study found that a positive long-term effect between TS and economic growth would be unlikely. Indeed, they indicate that the influence of a tourism-led industry on economic growth, while once viable at lower specialization levels, would fade at higher levels. Moreover, they suggest that lower levels of specialization may incur costs that, over time, increase. While at the onset of the industry, advantages such as a pristine environment and exotic atmosphere likely exist, there is risk and cost in maintaining the same benefits to which tourists were initially drawn. There arise penalties of environmental damage, needs for technological advances, safety issues, etc., requiring higher levels of specialization, all of which become demands of the tourists, all of which are costly.

Further, given the volatile nature of tourism, the ability to meet such costs becomes a risky endeavor as periods of dwindling returns negatively impact economic growth. Adamou and Clerides reveal a point at which tourism receipts ceiling, i.e., when tourism receipts plateau at 20% GDP. Therefore, their study contends that it is possible that TS no longer contributes positively to economic growth in the long run.

However, problems arise in Adamou and Clerides' study. They include the test protocol used (or not used) viewing growth rate perspectives according to tourism receipts. Sample dependence may have influenced their results linking tourism and economic growth. Their panel study combined 162 large and small countries, which may have reduced an accurate and meticulous view of the results and claims. Assuming that tourism receipts would support an affirmative bearing on increasing income would be a problematic contention insofar as consideration for the possibility of endogeneity was absent in the estimation of their fixed effects. Arguably, TS and economic growth may have feedback effects. For example, TS may increase economic growth, as indicated previously through the TLGH. However, economic growth can also boost TS by spending more on marketing and investing in tourism amenities (e.g., hotels, restaurants, and other recreational facilities) and public infrastructure.[13]

The third group proffers the idea that tourism incurs the Dutch disease. The Dutch disease is an economic *illness* that happens when there is an increasing amount of foreign exchange that generates a shift of reallocation of resources away from the original economic sectors to those benefiting from the boom in foreign exchange. This group of studies faults TS regarding its impact on the sectorial levels of the economy because TS modifies the reallocation of resources benefiting the non-tradable sector (labor-intensive) and impairs the tradable (manufacturing) sector (capital-intensive). Tourism-based economies prompt this sectorial shift because of a reduction in capital accumulation on the capital-intensive sector by increased capital and labor resources in the tourism sector. Also, tourism sparks an appreciation of the real exchange rate, which erodes the other export sectors' competitiveness. Ultimately, the benefits accrued by TS in the short and medium-term may shrink economic output in the long run or undermine the traditional sectors, including agriculture, energy, mining, and industry.[14]

---

[13] For example, Ridderstaat et al. (2013) found a bi-directional relationship between tourism and economic growth. See also Apergis and Payne (2012).
[14] See, for example, Copeland (1991); Chao et al. (2006).

The argument goes as follows: the destination is subject to higher prices and wages due to increased tourism demand and higher real exchange rate, which leads to inflationary pressures eroding the destination competitiveness. Since a significant surge in inward tourism flows tends to increase demand/consumption for non-tradable goods, the shift of domestic production factors away from the tradable sector may lead to a contraction of the industrial sector.[15] Some studies that examined TS in small islands found traces of Dutch disease. For example, Capo, Riera, and Nadal traced the illness in the Balearic and Canary Islands. Pratt also found signs of potential Dutch disease in his study regarding tourism's economic contribution in seven small islands (American Samoa, Aruba, Fiji, Jamaica, Maldives, Mauritius, and Seychelles).[16] However, in his previously mentioned study, Holzner contested Dutch disease's existence due to TS. His research found no symptoms of what he called *Beach disease* due to TS.[17] Inchausti-Sintes' study also concludes that, while tourism generated Dutch disease symptoms in the Spanish case, these symptoms are not especially harmful and can be mitigated.[18]

The TLGH can give rise to foggy, empirical analysis as is revealed when theoretical direction about the growth factors used in the model is loose, indefinite, or absent: the consequence of which could result in mercurial findings between studies. Exclusions or confusion between endogenous and exogenous variables in the regression phase can occur and are noteworthy. For example, addressing Dutch disease resulting from TS is the omission of the interdependence of tourists' consumption decisions with domestic consumers. Tourists' choices at the destination interfere with local consumers purchasing non-tradable and tradable goods and services, impacting local consumers' opportunities and choices. Tourism demand affects relative domestic prices, and their preferences affect the reallocation of resources (labor and capital). The presence of tourists at a destination is mainly due to the attraction of the destination (e.g., accessibility, heritage tourism, eco-tourism, gastronomic tourism, beach tourism, service quality), and probably their presence reveals the preference they attach to non-tradable goods and services over tradable goods.

---

[15] See, for example, Copeland (1991).

[16] See Capo, Riera, and Nadal (2007); and Pratt (2015).

[17] Inchausti-Sintes (2015) examined the potential of Dutch disease in Spain. They concluded that while tourism sparks appreciation of the real exchange rate, the positive consequences of tourism specialization could compensate for Dutch disease's adverse effects.

[18] Inchausti-Sintes (2015).

This preference may distort the economy's sectoral composition because of the competition for resources between, for instance, agriculture, and tourism. For example, land may be reallocated from low productivity activities, including agriculture and forestry, to tourism. In small islands, agricultural jobs are generally among the least protected, poorly remunerated, most hazardous, and of low status. That said, TS might affect the economy's sectoral composition without affecting its long run growth. Finally, it can be maintained that in circumstances where a country's tourism industry boasts more rapid tourist growth, the development and production of tourism activities race upwards as well.

Moreover, industrial strategies, procedures, and actions may also be an impactful force in a precipitous upward movement. For example, my studies assessing Aruba's case indicate that Aruba's TS was not an organic development. Instead, it was an intentional industrial policy to promote economic growth and development. Clancy also chronicled how Mexico, through an activist state, developed from the 1960s onwards five new resort clusters in Cancun, Ixtapa, Los Cabos, Loreto, and Huatulco. The government built these tourism clusters practically from the ground up through state agencies, including the tourism ministry (SECTUR), primarily through a national tourism development trust fund (INFRATUR and later FONATUR).[19]

7.4   HUMAN CAPITAL AND THE RESOURCE SCARCITY MODEL

Theoretical predictions are harsh to small island destinations. Predictions such as small islands cannot grow because they are small; if they engage with tourism, mainstream theory predicts that growth is elusive. The claim goes that economic growth will be confined to the early stages of TS-dissipating over time. Consider a small island relying on natural resources to stoke tourism development aimed at economic growth. Theory and empirical research coincide that TS embedded in natural resources will be short-lived.[20] TS is necessary, but not sufficient for economic growth. Alternatively, consider the other possibility: using technology to

---

[19] See Clancy (1999).
[20] Similarly, the literature also shows that investment is not the key to growth; it is necessary, but not sufficient. Easterly (2002) gives the example of the pancake recipe suggesting that increasing one recipe component (investment) will mess with the intended pancake. Similarly, foreign aid application on the premise of an investment/savings deficit has not pushed developing countries into prosperity.

spur TS for economic growth purposes. Technology is a critical driver of productivity, and productivity is the basis of prosperity and economic growth. Research shows that the hospitality industry (e.g., hotels, and restaurants) is significantly lower in technology content than in other sectors in the economy. This productivity gap is due to the lower impact of technological progress compared to, for example, manufacturing.[21]

By now, it is clear that the tourism industry does not seem to have the prowess that the manufacturing industry does as an engine of growth. The reason is that tourism cannot generate the same productivity level as the manufacturing industry. Chapter 5 already suggested, following Kaldor, that manufacturing is the engine of growth. The reason is, according to the neoclassical growth model predictions, that manufacturing has a high-level of technology content, which makes the manufacturing sector the primary driver of sustained economic growth. However, technology is a factor beyond the control of the economy, and therefore economic forces cannot explain its growth rate. That technology anchors on non-economic forces were the view of Solow, who anchors technology as exogenous. Non-economic forces determine technical progress by his account.[22] According to Solow's model, trade plays an ephemeral role to account for growth differentials in the long run. However, this progress does not work across countries. It would be a dangerous proposition to rely on something outside of your own doing or choice to prompt your growth. This condition brings us back to the issue of what determines the growth rate of the economy. So, what does this Solow claim leave us about TS's prowess to stoke economic growth?

Alternatively, the new growth models proffer an opportunity for trade to play a role in triggering economic growth. Exports open up opportunities for increased specialization, which leads to higher productivity growth through learning by doing. Human capital and its accumulation seem a way to keep the economy going, according to Lucas.[23] According to the Lucas model, it is human capital, and humans' ability to allocate their life choices in keeping with their values, including work, education, etc., that perpetuates and continues the means for economic growth to persist over time. Tourism research on the relationship between economic growth and

---

[21] See, for example, Smeral (2012, 2016); and Keller and Bieger (2007).

[22] Solow claims that machinery or capital cannot sustain growth in the long run, but technological change. The latter will economize on the fixed component, i.e., labor, avoiding diminishing returns. See Solow (1957).

[23] See Lucas (1988).

specialization reveals the human capital component as an indispensable affiliate for the tourism industry's economics.[24] The Lucas model questions the validity of long run economic growth anchored on exogenous factors. When applied to manufacture and tourism, the Lucas model predicts and shows that labor productivity in manufacturing is higher than tourism. This theoretical and empirical claim concludes that a tourism economy would experience a lower growth rate, and its growth potential is limited. The reason is that the manufacturing sector's accumulated knowledge cannot be fully transferred to the tourism sector because these two sectors are different. The logical conclusion is that tourism cannot grow via knowledge transfer from the manufacturing sector.

From the perspective of knowledge transfer's impossibility, TS would not be enough to sustain prosperity in small island economies. Or is it? At this juncture, it is critical to inquire if human capital is a growth-enhancing factor through TS and to discover the channel through which this growth takes place. Lanza and Pigliaru posit in a two-sector economy comprising of tourism and manufacturing that as long as these two sectors are substitutes ($\sigma < 1$), it suggests that tourism can grow the economy based on a demand approach. This demand approach is reflected in the destination's terms-of-trade performance. Marsiglio claims that this ToT performance is a choice made by the destination because residents' affecting endogenously determines the number of tourists visiting a destination about the physical and human capital accumulation. In other words, residents how much to invest in tourism facilities (e.g., how many hotel rooms) and how much to invest in education and training. As long as TS sparks a decrease in the proportion of physical capital in output generation and shifts facilitate human capital accumulation, TS would trigger economic growth.[25]

A second condition that may trigger relative prices favoring tourism products is through the application of scarcity (the resource scarcity model). We live in an era where global economies shift from a supply-constrained economy to a demand constrained economy. The enormous productivity gains that ensued since the Industrial Revolution resulted in more leisure time and consumption. In 1930, John Maynard Keynes predicted that we would have more leisure time available due to productivity gains, and he worried about what society meaningfully would do with this abundant

---

[24] See, for example, the special issue of Tourism Economics in 2007, number 4.
[25] See Marsiglio (2018).

time.[26] We know that society translated that much leisure time into more consumption-consumption in tourism and entertainment-consuming experiences and events. The economic problem no longer seems to be how much to produce, but how much to consume. We have arrived at an era of conspicuous consumption, long ago predicted by Veblen.

Conspicuous consumption goes beyond the mere satisfaction of the human need. Consumption embraces what it means to attain social status, a race for distinction through tangible or intangible possessions. This race for distinction consists of escalating consumption standards beyond the basic needs buttressing our relative needs, which we feel "only if their satisfaction lifts us above, makes us feel superior to, our fellows." Emotions (desires, wishes, wants, yearnings, longings, and preferences) dominate this subjective reality changing the meaning of scarcity from a function of natural resources (a lack of natural resources) to a social creation, a social choice! In the era of conspicuous consumption, the demand for goods and services is kept permanently scarce because goods, services, and experiences must be scarce for them to have a status, a distinction. Emotional scarcity is the use of a created, reduced opportunity to realize an experience or to obtain a service or product. The price that people are willing to pay derives directly from comparing that experience to other experiences in the utility of distinction. The limited opportunity to obtain that experience sparks more social status, and getting high social status means higher prices.

There are other upsides in practicing scarcity. The principle of scarcity could also induce a more efficient tourism practice. Because scarcity generates higher prices, business firms are motivated to economize on those experiences and services. In doing so, they minimize their costs and release factors of greater relative scarcity to those firms that value them the most, thereby creating more efficiency in the economy: and, hence, higher growth. In today's somewhat mutable tourism practices, we have already noticed a substantial increase in the influence of travel decisions motivated by scarcity. Another application of the scarcity principle is to protect the destination's reputation and maintain its quality as a product. Research suggests that reputation can overcome the moral hazard problems. Scarcity may induce higher prices, and higher prices can prevent quality deterioration. In his seminal study in 1997, Keane contends that firms are

---

[26] See Keynes (1932).

incentivized to cheat because of the transient nature of tourism consumption in a typical fragmented tourism system. He showed that repeat tourists' presence is the key to sustaining quality by securing premium prices that mitigate against the likelihood of quality deterioration.

A final justification for applying the resource scarcity model is because scarcity meshes with the small island reality. Resource availability is limited, such as land availability, which means that a small island cannot expand its tourism supply stock (e.g., hotel rooms) forever. Tourism supply in small islands is inelastic to demand, for example, due to land availability. I already alluded to scale issues that affect the adoption of productivity advances in small islands. Supply inelasticity in small islands allows extracting more economic rents from tourism prompted by scarcity. Extracting higher rents, however, requires quality service and a unique tourism-based environment. Thus, a small island can overcome the productivity deficit induced by tourism development through the lower opportunity cost of specialization. The principle of scarcity in the tourism supply can offset differences in physical productivity. Charging premium prices embedded in scarcity and quality is the foundation of a competitive destination. The opportunity to capture premium prices hinges upon understanding what drives social status. This understanding involves higher levels of human capital to create and organize opportunities for conspicuous consumption. It is a question of effective destination management to drive ToT.

7.5   AN EMPIRICAL VALIDATION

Effective management requires informed decisions. Informed decisions rely on improved knowledge of the factors contributing to the influx of tourist arrivals, and tourism receipts are needed in small islands. Moreover, the question is whether the information afforded via empirical study runs true in the short-term and the long run. Because the interdependence between TS, tourism-related business development, and the benefits of both sharply reverberate through the destination's economic stability and its people, the need for critical and empirical analyzes of tourism demand is required. However, while benefits have been effectively ascertained in the short-term, issues on a global stage (e.g., political, business cycles) and natural and other characteristic elements of the destination (e.g., weather) leave long run assumptions uncertain.

That said, because small island destinations' survival could depend on the tourism industry's growth and sustained status, they can little afford to rely on non-credible or unverified information to strategize their primary economic industry's growth. Without valid information as to demand, the risk of failure is imminent. However, as stated, empirical, and practical information about market demand as it impacts the production of goods and services would bolster the much-needed control and growth of the tourism industry in small island destinations. If future resource requirements for tourism are estimated accurately and provided to formulate national development plans, a useful measure of the demand for tourism goods and services must be available. Small states and developing countries seeking to make tourism a viable component of their development strategy could benefit enormously from meaningful international demand estimates.

The value of empirical study and data analyses of tourism development to respond to small islands' scale distresses are an indispensable tool to offset and control such distress. Small islands face population build-ups resulting in pressing socio-economic challenges, such as employment and social stability. Thus, it becomes crucial for small islands to effectively strengthen their market base and encourage and capture new spending markets that would access the destination. The result would allow for the continued emergence of businesses to meet the demands of tourists. Opportunities for employment and additional strategized development to complement and supplement tourism needs would cyclically flow by nature to balance the industry with the goods and services afforded by tourist demand. However, without scientific evidence of how demand impacts the economic structure of tourism and thus the ability of sound decision-making capacities necessary to forward the industry, it is likely that the small island destination could relegate itself to a guess and check the type of analysis-the like of which would indeed be too arbitrary to be an adequate response to a problem as profound as poverty.

The validation of the conditions for growth is critical to the choice of specializing in tourism. As noted previously, any choice entails resource allocation, and resource allocation is a hard and costly task. Making a choice is incredibly hard and stressful, albeit crucial, for a small island because of its limited development opportunities. Insights as to the impact of volatility on economic growth, the entanglement possibility between TS and the Dutch disease phenomenon, and the permanence of economic growth due to TS are critical issues for small island development. Unfortunately, many countries still give a low priority to tourism information based on scientific research. If small islands ignore the importance of scientific data and rigorous analysis, they will soon find themselves unable to compete effectively in tourism's global marketplace. While evidence indicates that technological progress or innovation is the backbone of nations' wealth, relying on tourism development for economic growth and prosperity masks an inherent dilemma: the tourism industry is the least productive sector compared to manufacturing, or is incredibly hard to measure.[27] Due to increased competition, mature destinations reach the point of exhaustion of existing resources and erode their rationalization potential.

The prior discussion requires some empirical embeddedness to assess the impact of ToT on a small island's economic performance. The empirical assessment included data about ToT, TS, and GDP per capita data from 18 small islands across four continents (The Americas, Europe, Africa, and the Pacific).[28] The data cover the years 2000 to 2018, including two critical shocks, i.e., September 11 (2001) and the Great Recession (2007–2009). The data for the variables were obtained from publicly available World Bank, World Tourism Organization (WTO), and countries' Central Banks and Central Bureaus of Statistics databases. The small islands sample accounted for history, geography, endowments, economic structure, tourism development, and development level. The first group of islands is located in the Americas: Antigua and Barbuda, Aruba, The Bahamas, Barbados, Dominica, Grenada, St. Kitts and Nevis, St. Lucia, St. Vincent and the Grenadines, and Trinidad and Tobago. The second group of islands is located in Europe: Malta and Cyprus. The third group is from Africa: Cape Verde, Mauritius, and Seychelles. Moreover, the fourth group is from the Pacific: Maldives, Vanuatu, and Samoa. The panel is strongly balanced, including data for all countries and years.

The variables entered the model in first differences, based on the Im-Pesaran-Shin (2003) panel unit root test, to ensure stationarity. The lag order (lag=1) is selected based on SBIC, AIC, and HQ criteria. First difference suggests that ToT has a permanent effect on the future progress

---

[27] Productivity is hard to measure in many service sectors. Consider health care where data show little improvement in TFP despite dramatic improvement in life expectancy and well-being.
[28] The analysis applied the structured, focused comparison of Alexander George to determine the sample selection.

of economic growth. The Westerlund test for cointegration indicates that the three variables are cointegrated at 1% (statistics = 8.2527; p = 0.0000), which means they move together over time. Next, the endogeneity test based on the Davidson-MacKinnon (1993) test of exogeneity of TS is exogenous (F (2281) = 11.8160 and p = 0.0000) is firmly rejected, suggesting TS could be endogenous, which confirms the necessity of employing the IV regression method.

TS and GDP per capita may have feedback effects revealing endogenous dimensions. For example, tourism marketing can be increased by more tourist spending, or the quality of tourist services can be improved through investment sparked by higher tourist spending. Better quality of tourist services and more marketing efforts can boost the arrival and spending of tourists. Tourist demand obeys to forces beyond the small island (income elasticity in the source country). However, demand also depends on the quality of the tourism product (destination) that is endogenously determined.

The analysis also included a systematic examination of the relationship ToT, TS, and GDP per capita growth. The model regressed the average GDP per capita growth rates on the ToT natural logarithm, TS, and TS-square to justify potential diminishing returns.[29] The model accounted for these potential endogeneity effects employing the IV regression approach. The model also included two dummy variables covering September 11 and the Great Recession. The fixed-effects model turned out to be the appropriate test to run based on the Hausman test. The data reveals that the countries entertain (between effects) differences while each country exhibits differences over time (within effects). The results show significance at the 5% level for ToT, and significance at the 1% level for TS, all with the expected signs (see Table 7.1). The results also reveal that TS has a standardized beta coefficient with a value of 0.975 and ToT has a beta of 0.061. The model shows that with every increase of one standard deviation in TS, GDP per capita rises by 0.975 standard deviations. This assumes that the other variables are held constant. With an increase of one standard deviation in TS level, GDP per capita rises 0.975 standard deviations—again assuming all other variables are held constant. The results suggest that the ToT effect on GDP per capita is substantial, almost a one-to-one relationship.

---

[29] Stochastic and dynamic optimization (inertia and price stickiness) could prompt a behavior that is non-quadratic.

TABLE 7.1   Panel Results Tourism Specialization and Terms-of-Trade

	Dependent Variables: LGDPCAP	
Independent Variables	**Unstandardized**	**Standardized**
Tourism specialization	5.037***	0.975***
Tourism specialization-Sq	−4.452**	−0.631**
Terms of trade	0.295**	0.061**
Sept eleven	−0.409***	−0.158***
Great recession	0.084**	0.038**
Constant	6.941***	0.035**
Rho	0.939	0.939

Notes: ***indicate significance at 1%; **indicate significance at 5%.

Let us unpack these results. First, the sign governing the relationship between ToT and GDP per capita suggests that they complement each other. The positive sign of ToT on GDP per capita also suggests that the resource curse exposed to economic development does not exist in the sample of islands. Or, if it does exist, its impact seems marginal.[30] The effects of ToT on GDP per capita also entertains long-term effects. These three aspects, complementarity, the absence of the resource curse aspects, and the long-term effects of ToT combine as a germane channel for sustaining economic growth. Second, TS and GDP per capita cultivate a long-term positive relationship, consistent with a growing consensus in the tourism literature embracing TLGH. This positive long-term relationship suggests that TS is a crucial source for small islands' economic growth and development, and in overcoming scale and the productivity gap.

Third, the findings, however, also indicate that TS entertains diminishing returns, which means that the effects of TS on economic growth may be slowing down over time, and that the positive effects of specialization on economic growth could potentially disappear altogether. The tourism literature interprets diminishing returns due to TS as a disease symptom that would hamper output growth over time. In a recent study that I and two other UCF researchers conducted, we showed that Malta's tourism development was showing ominous signs of diminishing returns and was in desperate need of restructuring its product.[31] Discovering a tourism inflection point

---

[30] For a discussion on the resource curse, see, for example, Inchausti-Sintes (2015); Capo et al. (2007).
[31] See Croes et al. (2018).

or threshold does not mean that the tourism industry is doomed to failure or has converted itself into a land mine destined to implode the small island. The Balearic Islands were able to parse and to surpass their inflection point for a stronger tourism development performance. In their interesting article, Aguilo, Alegre, and Sard discussed how the Balearic Islands were able to restructure their tourism product and to competitively reposition the destination.[32] Marsiglio also provides the theoretical underpinnings of how a small island destination can reposition itself with stronger and long-term TS effects on GDP output.[33]

Fourth, a paradox emerges between the long-term positive effects and the diminishing effects of TS on GDP per capita. To compare their relative importance, I standardized these variables, which means that they can be compared to each other in the model. When comparing the standardized coefficients of TS (standardized coefficient=0.975; p=0.001) with TS-square (standardized coefficient = –0.631; p = 0.012), the picture clarifies. The larger standardized coefficient of TS compared to the TS-square suggests that the impact of TS on GDP per capita will remain positive. The TS trend counterbalances the negative cycle or volatility by 34.4%. To say it differently, an increase of one standard deviation in TS level, the net GDP per capita rises by 34.4% standard deviations — again assuming all other variables are held constant.[34] It appears that TS develops channels that stimulate human capital accumulation over time, becoming the growth engine.

Fifth, while the long-term TS effects are stronger than the diminishing effects, a potential new picture is emerging in which TS may lead the small island destination to a low productivity growth path. This journey may reveal a steep transitory growth path that stabilizes over time, exposing lower economic growth. This hump-shaped relationship between TS and GDP per capita depends on whether physical and human capital is complementary or are substitutes. The previous results suggest that, despite the presence of volatility, TS nurtures an upward trend in the GDP per capita in the sample of small island destinations.

Another critical aspect of TS on GDP per capita is its susceptibility to volatility from external shocks besides the interplay of physical and

---

[32] See Aguilo et al. (2005).
[33] See Marsiglio (2018).
[34] Arezki et al. (2009) found that that one TS standard deviation would lead to 0.5 percentage point per year in growth, ceteris paribus.

human capital. TS enables small islands to take advantage of the global market (openness). Terms-of-trade (ToT) captures this advantage from openness. The previous findings already indicated that ToT has a positive impact on output growth. However, its positive effects may be undercut by volatility, which is the costs induced by openness. Volatility refers to the fluctuations or shocks in the ToT. To measure these disturbances' impact, it is necessary to decompose the ToT in its secular trend and variations or cycles around the trend. Cycles measure volatility. One way to pick up volatility is to decompose ToT through the Hodrick-Prescott (HP) (the applied λ value=6.25) filter to decompose the ToT into trend and cycles. The identifying assumption is that ToT shocks are exogenous. Consider the following: business cycles in source countries may impact the volatility level of tourism demand flows with damaging consequences in service quality, market opportunities, profits, jobs, and welfare. The Great Recession in 2007 stalled tourist arrivals, deteriorated economic imbalances, and sparked declining room capacity, and fewer flights to the Caribbean region. It took nearly two years for Caribbean destinations to bounce back from the recession.

The impact of the ToT trend and volatility are presented in Table 7.2. The trend is the long run perspective of the relationship between ToT and economic growth, while the cycle depicts the short run perspective. The standardized coefficients measure the impact for comparison purposes— changes in the ToT trend positively associated changes in GDP per Capita at the 5% significance level. The result confirms the TS hypothesis anchored in the positive relationship between terms-of-trade and economic growth. The ToT cycle had the correct negative impact on growth; however, its impact is not significant. Even if cycles were significant, the long run positive ToT trend (Beta coefficient= 0.092) more than compensates for the short run negative volatility changes (Beta coefficient = –0.006). The positive sign of the ToT trend seems to undo the negative cycle effect suggesting that TS (openness) could export a larger share of output. The ambiguity of cycle effects on GDP per capita indicates that TS volatility does not seem to hinder positive economic output, which questions the stream's critical literature on TS.

The intensity and magnitude of the interplay between physical and human capital magnitude triggered by TS on GDP per capita is a choice made by the destination's residents. This choice stems from endogenous factors as indicated in Chapter 2, because residents ultimately decide how many hotel rooms and land availability are allocated to tourism facilities and amenities. They decide how much they invest in generating competencies and skills. And they decide on how many efforts to engage in coordination in tourism production. Ultimately, residents decide how many tourists will visit their small island based on their allocation of money, time, and efforts in making the destination attractive to tourists with the willingness to pay for their products. Arguably, the volatility stemming from tourists' willingness to pay is outside of residents' control. However, as the previous empirical exercise and validation demonstrated, this volatility can be counterbalanced by the appeal and attractiveness of the tourism product, which lies in the destination residents' hands.[35]

TABLE 7.2   Panel Results Tourism Specialization and ToT Trend and Cycle

Dependent Variables: LGDPCAP		
Independent Variables	**Unstandardized**	**Standardized**
Tourism specialization	4.947***	0.957***
Tourism specialization-Sq	–4.273**	–0.252**
Terms of trade trend	0.463**	0.037**
Terms of trade cycle	–0.126 (NS)	–0.016 (NS)
Sept eleven	–0.409***	–0.021***
Great recession	0.085**	0.016**
Constant	6.175***	0.016**
Rho	0.939	0.939

Notes: ***indicate significance at 1%; **indicate significance at 5%; NS indicate non-significance.

Of course, the effects of these two aspects differ across small island destinations impacting the magnitude of economic growth rates. Island specific effects, such as management skills, institutional strength, coordination prowess, political will and vision, and social cohesion, seem to explain the variation in performance. The results manifested by Rho=0.939 indicate that some small islands perform better than others. The Rho result is consistent with the broader impact of the between—country effects compared to the within—country effects. Some islands reveal better

---

[35] Economic agents must cope with randomness and non-linear effects prompted by bounded rationality of others. An illustration of randomness and non-linear effects are inertia and price stickiness that may impact the short-run relationship between ToT and economic growth.

economic outcomes than others. The results reflect the kaleidoscopic reality of small islands, suggesting that the relationship between TS and economic growth is not self-evident and automatic. Other relevant factors seem to be related to management capabilities, resource endowment, human capital, institutional strength, specialization, and residents' attitude towards tourism development, and the choice to specialize. Indeed, the market should allow these resources to go to their most valuable use without distortion, which remains hard in a small island destination due to scale challenges.[36]

While the results suggest that TS has a declining effect on economic growth, it still has the prowess to trigger economic growth. This growth is due to the stimulation of human capital accumulation, which is the source of the dynamics in services innovation. One of these innovations is the co-production or co-creation thesis that underscores the high degree of interaction between tourists and tourist suppliers. This interaction is at the heart of tourism organizational innovation, market innovations, and network innovations. Tourism production may entail higher costs over time because of its inherent lower productivity, and it is hard to scale up. But small island destinations are in good company because history shows that developed countries have embraced services by timidly becoming a service economy. It seems as if shifting resources towards parts of the economy where productivity growth is lowest appears to generate more value.

A plausible explanation is that people or tourists desire services that are personal and of higher quality. Empirical evidence is showing that while the cost is important, the cost is not everything. Tourists want better quality service and products. As tourists demand more unique experiences to from which to choose, the interaction with the local population becomes of strategic management importance. The interaction between the tourists and residents could shape, define, and infuse meaning into these unique experiences. The quality of that interaction could be a powerful and intangible attribute of the destination and could determine the impact of TS on growth and well-being. This impact is not direct but relies on many channels and factors, such as residents' willingness to interact with tourists, the economic benefits accrued to residents, types of jobs, the quality and strengths of the institutions, the degree of commitment of the political, economic, and social leadership, and the impact of tourism on

---

[36] Banerjee and Duflo (2019) identified the misallocation of resources as the major growth impediment in developing countries.

residents' well-being. There is a great need for an in-depth understanding of the risks and the means of managing human agency, processes, and productive activities related to the tourists and residents' well-being.

7.6   A LOGISTIC DEMAND FUNCTION

Staying attuned to the evolving tourists' preferential scheme might entail a logistic demand function's characteristic shape. Demand for the tourist product may show lower demand elasticity over time, suggesting diminishing marginal returns beyond a certain income threshold in the source country. This diminishing return possibility is why it is so essential to examine the segmentation potentials in rich countries. Alternatively, the supply side of tourism also reveals an attractive feature. The tourism industry supply, unlike supply in the manufacturing industry, does not always follow demand. Typically, tourism supply shows low growth and is a result of either high demand or limited supply. Natural resources are scarce, and congestion may make the destination less attractive. The existing supply constraints in small island destinations are land availability, accommodation capacity, and environmental resources.

Tourism demand provokes direct, indirect and induced effects. Tourists when visiting a destination spend money on accommodation, restaurants, activities and attractions, and entertainment. These spending constitutes their direct spending which extends to firms and businesses, households and government. These economic agents on their turn buy goods and services from other economic sectors. Indirect spending emerges from public and private investments and the impact of purchases from suppliers and vendors. Finally, economic activity also stems from household spending of income directly or indirectly earned in the tourism sector when purchasing shelter, food, clothes, entertainment, and services.

How tourism money is allocated within an economy is determined when tourism-related businesses acquiesce to tourists' demands. The tourism industry is driven by the ever-present need to ascertain and assimilate the wants and needs of the tourist who seeks to experience a memorable involvement with the destination. Thus, the successful allocation of tourism money is best entrusted to savvy businesses in appropriating a sufficient tourist market share opportunistically. Moreover, these businesses are efficient in determining and developing the economic growth opportunities that services linked to tourism could bring (such as finances, insured safety nets for tourists, products to fulfill tourist essentials and demands, etc.). It then becomes possible that a chain reaction of expectations could occur in the increased quality of entrepreneurial offerings. These expectations emerge because destination entrepreneurs must continue to produce the kinds of quality products offered by global marketers- or risk a backward movement in their quest for tourism-led growth (TLG).

Combining scarcity, quality, and the right tourist profile (from income elastic source countries) requires a talented human capital. Because tourism development production becomes more expensive over time due to the slow advancement in productivity, production costs increase faster than other economic sectors. Tourist offerings and services are required to entertain higher prices to compensate for higher production costs. Nevertheless, higher prices are only possible through high-quality tourist offerings, which require the industry's capacity and capability to deliver new and innovative products and services. High competencies, skills, and creativity from human resources are crucial ingredients undergirding the tourism sector's future sustainability in small island destinations.

While the previous regression exercise cannot say why countries differ in their performance, I can surmise from research that human capital may have a lot to do with the last empirical exercise results. Elsewhere, my research has shown that human capital investment is required to deliver quality offerings and services to tourists. What seems at stake is how tourism destinations are managed and how they assemble and make their choices to raise the quality of life of the average person. Globally, the competition for market tourist shares has loomed. As a result, destinations exist in a chronic state of shifting policies, principles, products, and procedures to hedge their market share while edging out the competition. As each destination struggles to maintain those hedges, the quality of their goods and services has, by necessity, increased-, and come to be expected by the tourists. The general interpretation of the previous empirical exercise is that higher human capital could usher the second phase of TS of continued economic growth of small island destinations, pushing the production frontier to the right to benefit the residents' well-being. Pushing the frontier to right requires an intentional government policy and incentives and implicates those destinations may continue to grow just by allocating their existing resources to more appropriate uses.

In the next chapter, I will wrap up our TS thesis and discuss the need for a learning environment to support informed choices to allocate resources to their most valuable use. I will conclude with the impact of COVID on small island destinations.

KEYWORDS

- economic growth
- gross domestic product
- led-growth strategy
- logistic demand function
- tourism specialization
- tourism-led growth hypothesis
- World Tourism Organization

CHAPTER 8

UPENDING SCALE

In my 2006 study and 2011 book on the *Small Island Paradox*, I asserted that tourism specialization (TS) is a vital dimension to the survival of small islands because it provides three vital components germane to the cultivation of development opportunities, i.e., additional demand, efficiency, and cost discovery. These opportunities prompt economies of scale for more goods and services by decreasing production costs and increasing competition among local firms.[1] I additionally argued that greater openness triggered by TS could also generate the opportunity for local entrepreneurs to quickly engage in cost-effective learning of the international market's taste and preferences without venturing outside the domestic market.

8.1 THE POWER OF TOURISM SPECIALIZATION (TS)

Consider, for example, that destinations attract tourists that are keen on determining the fulfillment of the wants and needs that ensure a positive experience for them. However, that determination comes at a costly price generated by research, data collection, analysis, availability of resources to address tourist preferences, and coordination skills, even perilous trial, and error. However, if local businesses could take advantage of the international information already positioned by specialization's greater openness, then information-seeking costs to the local entrepreneur could become reduced and manageable. The product information that is already available, its quality, and branding positioned by the greater openness of specialization situates successful exportation as conceivable. The result may be more significant economic linkages and a more robust and diversified economy. Moreover, such proliferate circumstances bode well for the small island

---

[1] See Croes (2006, 2011).

as it strives for a competitive place among other desirable destinations. Further, rival competition should help the strong strategic focus of the small island. Competition increases efficiency via the allocation of inputs leading to an increase in total factor productivity. It also triggers better management practices and systems of organization, more training, and increased capacity utilization. These positives also have progressive effects on other economic sectors due to spillover and spillback standards, affecting total factor productivity. The positive confluence of these effects then defines how attractive the destination is for international tourists, impacting the international arrivals and spending numbers.

In constrained, riddled environments, such as small islands, tackling their economic organizational problem is a must. Small island problems can arise from various centers whether such centers are a consequence of economic difficulties, the environment constrained by lack of resources, or problems organizing and incentivizing the kinds of enthusiastic and thought-provoking human behaviors necessary to contribute to productivity. The issue borders on harvesting models that reveal coordination of collective societal influences that include benchmarks to determine the worth and usefulness of those influences. Alas, small islands experience biting problems resulting from their difficulties in synchronizing models and human behaviors that would comprise or contribute to positive productivity.

Many small countries embraced tourism specialization (TS) as a suitable development strategy. In particular, small islands seem prone to engage in TS because they entertain a smaller opportunity cost. A popular proposition among social scientists and international organizations was the idea that small islands could thrive, prosper, and provide a decent quality of life to their residents by applying TS as a pathway.

My own experience shows how small islands can overcome dire situations despite their smallness and become economically successful. As a native of Aruba, a small island in the Caribbean with a population of about 100,000, I experienced first-hand how global forces have jarred that small island in the early 1980s and why searching for alternative economic opportunities was so daunting. Economic opportunities were naturally scarce. Furthermore, taking risks in any economic venture was confronted with the standard claim that small islands' vulnerabilities revealed uncertain prospects. Vulnerabilities are associated with the risks and sensitivities of small island destinations to shocks and disturbances.

Mainstream tourism literature appends these shocks and disturbances to the pessimistic view regarding the viability of small islands. Because of globalization, impacts of shocks and disturbances can spread rapidly in unexpected ways (e.g., SARS, the global recession of 2008–2009, and hurricanes), causing havoc to the socio-economic fabrics of small island destinations. According to this view, the increased uncertainty of the impacts, their timing, and the speed with which they spread may trigger high levels of volatility in tourism demand, thus hurting these small island destinations.

What struck me then and now is this claim regarding lack of economic opportunities and that enduring vulnerability was premised on a passive or lack of creative agency. It is as if the creative agency of small islands' residents is framed, molded, and limited by its physical size. Aruba's 1980's jarring of the failed oil industry witnessed a need to access creative means to seek an alternative and viable economic opportunity. The resulting recovery of Aruba's economic viability and the outcome of that 1980's shock illuminated the creative resourcefulness and resilience of the Aruban people to overcome a potentially lethal economic disaster. Indeed, their creative success could not be consigned to the limits of the island's physical size: nor could their accomplishment be designated fortuitous and diminished by literary statements that dispatched creative agency as regulated or constrained by geographic size.

To more precisely explain the shock of Aruba's oil industry dissolution, consider that it was the economic mainstay for about 70 years. By 1984, it collapsed due to adverse coalescing global forces. Overnight, the island experienced an upward and dramatic jolt in its unemployment, reaching a devastating 25% of the working population-a number reminiscent of the U.S. Great Depression. The only real alternative to unemployment was tourism development. Thus, Aruba embarked on developing tourism quickly to resolve its vast unemployment post-collapse of the oil industry.[2] Could Aruba, which forcefully shifted its development strategy from the oil industry to tourism development, imbue meaning, and relevance to the questions I posed in the book preface and addressed throughout this book?

Switching from oil to tourism included a disadvantage because of the productivity gap inherent in tourism development. Tourism in small

---

[2] See, for example, Croes, R., (2010). *Anatomy of Demand in International Tourism: The Case of Aruba.* Lambert Academic Publishing: Saarbrucken, Germany; and Ridderstaat, J., (2007). *The Lago Story: The Compelling Story of an Oil Company on the Island of Aruba.* Oranjestad: Editorial Charuba.

islands mainly consists of small, labor-intensive businesses that exhibit difficulty in rationalizing, suffering, or recovering from the consequences of cost disease. As explained in previous chapters, cost disease refers to increases in labor costs in the tourism sector while lacking adequate productivity to offset that cost. This cost disease is due to the labor intense core of such businesses and their need to compete with other economic sectors on production factors (especially regarding the talents and skills of the workforce, which are scarce in a small island). These tourist businesses lack the productivity dimensions compared to other sectors such as the oil industry. Moreover, they must offset the higher labor costs with higher prices of their products and services, which could then render them uncompetitive in the international market. Further, this productivity gap may spawn high specialization costs. The need to overcome cost disease is, indeed, confounding.

However, in its early stages of redevelopment post-oil collapse, Aruba showed that specialization costs could be overcome. The case suggests that the propounded constraints suggested by small size were either not disabling or that Aruba beat such constraints. Indeed, Aruba was able to cope with the coordination problem so pervasive in economic development. The coordination problem is one of the major puzzles to solve in TS.[3] In the case of Aruba, it centered on resolving the sequencing issue of airlift and hotel rooms (what should come first) or investing in a new airport to include the United States pre-clearance system:[4] alternatively, whether to wait until the minimum amount of tourist's arrivals warranted the airport investment. Government's vision, intention, and steady involvement was critical in Aruba's pivoting to TS.[5]

The Hague was in the way of Aruba's ability to solve its coordination problems because The Hague's support was initially lacking in Aruba's efforts to convince the United States to establish the pre-clearance facility on the island. Because Aruba is a non-independent country within the Dutch Kingdom, any agreement with an international country would need to be ratified by The Hague. Moreover, negotiations leading to the treaty that would accomplish this were cumbersome, and the agreement between

---

[3] See, for example, Candela and Figini (2010).
[4] Aruba is one of the only two islands in the Caribbean with the U.S. preclearance system that includes the presence of U.S. immigration and customs, which means that all U.S. bound flights are domestic flights between Aruba and the United States.
[5] Mazzucato (2018) in her insightful book argues for government and public value because profits and value stem from collective efforts.

Aruba and the United States was achieved despite resistance from The Hague. I know because, as Lin-Manuel Miranda would say, *I was in the room where it happened.* Experimenting to solve coordination problems is risky because there is only one way to be perfect and a million ways to be imperfect and fail, thus the complex and cumbersome decisions and details of the treaty.

The Aruba case is the more surprising because the literature regarding tourism and small countries revealed, as a condition for the tourism-growth nexus, the need for a structural change shifting from low productivity (e.g., agriculture) to a high productivity sector (tourism). For example, Hernandez-Martin asserts this factor as the shaper of economic growth in small countries. However, Aruba's experience was different: the structural change occurred from a higher productivity sector (oil refining) to a lower productivity sector (tourism). The implication is that the productivity gap by itself may not be the *explicandum* of economic growth in small economies. This structural shift also questions the proposition that resources will flow towards high productivity parts of the economy. Arguably, economic value seems to rest in providing outstanding personalized service and quality products through collective efforts and coordination.[6]

Admittedly, experimentation, and creative solutions were not the sole purviews of Aruba. Instead, experimentation, and creative solution seem an intended fit for small islands. Baldacchino chronicles how small islands used their resourcefulness to realize achievements that are incommensurate with their small size. He refers to this resourcefulness as innovative governance. The manifestation of small islands' creativity is also present in the Canary Islands, which could wield relevant and meaningful socio-economic concessions from continental Europe to counteract socio-economic imbalances in continental Europe. The import tax (AIEM) and the offshore zone (called the Canary Islands ZEC Zone) are several of the economic-financial measures called the Economic and Fiscal Regime of the Canary Islands (REF) that protect in some way from imbalances with the continent.[7]

---

[6] As mentioned in previous chapters, this shift is known within economics as unbalanced growth where resources tend to shift towards where productivity growth is lowest, as is what happened in the United States where employment moved from manufacturing to service.

[7] The Canary Islands ZEC Zone only requires firms to pay 4% of their profits and encourages them to reinvest in their business development.

8.2   MEASURE WHAT COUNTS

However, the empirical analysis of the impact of TS on economic growth is confounding. Although tourism benefits to economic growth occur in the short-term, it is not necessarily true that these benefits would happen in the long run. The influence of tourism receipts on economic growth may not be as transparent or easily interpreted as would be desired. For the small island destination, this is especially true. So, while tourism has posed considerable economic involvement globally, its contribution to small island economic development has not been quickly or consistently evident. Whether the impact of TS on economic growth is sustainable is crucial, particularly for small islands in their quest to beat the scale challenge, cement prosperity, and enhance well-being. Inconsistent results in scientific inquiry are not new and are not harmful either. The essence of the scientific inquiry is to test the validity of assumptions and propositions precisely. In other words, it is making them falsifiable following Popperian precepts.

Nevertheless, tourism is a particularly interesting case in examining resource allocation because it entails the opportunity to influence price levels. In other words, a small island destination may have some price-maker power. The relevant questions are the conditions that generate these opportunities. As discussed in Chapter 2, the tourism demand level consists of two mutually reinforcing dynamics, i.e., the exogenous and endogenous dynamics. These two dynamics interact, and their result cannot be determined *a priori* but is the outcome of empirical validation. The tourism demand literature centered mainly on the exogenous dynamics, identifying income in the source country as the most vital factor. Tourism research indicates that international tourism is income elastic. As income in source countries increases, residents from these countries are more likely to spend a more significant portion of that increased income on vacation. Arguably, tourist receiving countries should only assess the increase in source countries' income levels to rely on TS. However, receiving countries in this view have no way to influence the source countries' income level. If this perspective is the only condition, then receiving tourist countries, including small islands, would be price-takers!

The other dynamic force comes from how residents in the receiving tourist country determine their development choices. This perspective argues that the tourism demand level is endogenously determined by the residents' choices regarding physical and human capital accumulation, and their ability to put resources to their most valuable use. Of course, tourism demand also obeys to the economic state in the source country beyond the small island (exogenously determined). However, ultimately it is the attraction of the destination that matters (endogenously determined). A small island attracts tourists based on its quality as a tourist destination. This quality references image, facilities, quality of hotels and services, environment, cultural heritage, political stability, and the human capital. The quality of tourism services depends on the investments made, and these investments depend on choices based on anticipation of the return (e.g., profits).

Choices depend on the human capital ratio to physical capital that prompts the dynamic economic intensity. How specialization promotes economic growth via tourism occurs through demand (income elasticities) and supply-side factors (learning-by-doing). We already elaborated on the critical role of learning in generating growth in Chapter 2. Tourists may change or adjust their preferences once they learn that their knowledge about a product is wrong and adjust by buying less or not buying. A tourist producer must also decide how much quantity to produce and its price given his financial situation. If the sales do not comport according to his expectations and leads to unintended consequences to his financial situation, he will promptly adjust. This acceptance and acquisition of this new reality (new knowledge through learning) are endogenous. The producer will learn with every decision that he takes, making the process irreversible. This process may engender long run change and may make the triggered production externalities permanent.

Production externalities' consequences can entertain positive or adverse trends, or slow down a small island's growth potential through diminishing returns. These consequences determine if the learning behavior is consistent with the acquired knowledge and whether it will move development to a new equilibrium. Whether the equilibrium is permanent or transitory shapes the intensity and magnitude of the relationship between TS and economic growth. The lack of consistency, for example, could include cycles that may deviate from the trend. Effective attention and concern for such information may facilitate a small island's understanding and manipulation of what a small island *is* producing and what it *can* produce. That said, it is about knowing the relationship between what a small island is producing and its production potential. This gap between actual and potential output may be a powerful way of assessing the growth potential of TS, its lingering effects and patterns, and its resiliency against unwanted exogenous shocks stemming from openness, natural disasters, and pandemics.[8]

8.3   THE POTENTIAL COVID IMPACT

Shocks to the tourism industry happen all the time. Such were hurricanes, tsunamis, SARS, MERS, EBOLA, bird flu, swine flu; and such shocks happen with uneven impacts across the globe. The previous chapter shows that 911's impact was significantly different for the sample of small islands compared to the Great Recession (see Tables 7.1 and 7.2). That is, while the former was negative and significant, the latter was positive but insignificant to small islands. All these shocks have a relatively short duration in common, and they did not halt international mobility.

COVID, as a shock, turned out to be very different. Its impact is dramatic because it halted international tourism to a complete standstill for a few months, burning economic growth twice the level of previous shocks and carrying the potential of sparking structural changes in tourism supply and demand. The World Tourism Organization (WTO) prognostication reveals that COVID would drop international tourist arrivals 20% to 30% and between US\$ 300–450 billion, equivalent to nearly one-third of the US\$ 1.5 trillion generated in 2019. According to the same WTO report claims, this loss's value is five to seven years of growth.[9] Examination of previous major viral epidemics suggests a 19 months average recovery period, while some destinations managed recovery in only 10 months.[10]

Admittedly, the most affected destinations are small islands because they rely more heavily on international tourism for their economic growth and prosperity. For example, travel, and tourism generate about US\$ 30 billion per year for SIDS islands. A decline of 25%, which equals to US\$ 7.5 billion, would spark a decline of nearly 8% in their total GDP average. The United Nations estimated that small islands' GDP would shrink by

---

[8] I will discuss the consequences of COVID-19 in the next sections of the book.
[9] Visit: https://www.unwto.org/news/international-tourism-arrivals-could-fall-in-2020.
[10] Visit WTCC: https://wttc.org/News-Article/Containing-the-spread-of-panic-is-as-important-as-stopping-the-coronavirus-itself-says-WTTC.

4.7% in 2020, compared to a global contraction of around 3%.[11] The toll is higher for tourist dependent economies, according to the IMF, predicting a contraction of more than 10% in their GDP.

Arguably, this average drop would be uneven across these islands. The negative impact on the Maldives would equal to a drop of 17% of its GDP, Grenada would lose 14% of its GDP, Fiji would lose 10%, and Barbados 9%, according to a recent UNCTAD report.[12] The Dutch Caribbean Islands of Aruba, Curacao, and St. Maarten are expected to waste between 14% and 18% of their GDP. The Balearic Islands may experience a 30% decline in its GDP.[13] Alternatively, Malta is projected to lose less than 3% of its GDP due to the pandemic.[14]

The pandemic is a supply shock affecting international tourism because countries closed their borders to control the virus spread.[15] Closing borders meant that islands were secluded from scale (openness), and the lack of accessibility affected their ability to tap into the global market. As their domestic market is small, promoting domestic tourism would have limited effects. Cloistered in their own space would cause backward leaps for the small islands' sustainability. Arguably, the pandemic reveals how vital and meaningful openness is for small islands. Yet, another important manifestation of small islands is that their economy did not collapse while vulnerable to external shocks such as COVID. As previously indicated, COVID impact is enormous, yet it does not affect these islands and their tourism industry in an existential way. One would expect that because of the heavy reliance of small islands on tourism and the halting of international mobility, these islands would go under and that many of their businesses would fail into bankruptcy. While a possibility, I have not noticed this frightful trend in small islands. The reason seems that again small islands are showing their resourcefulness in how to get out of the pandemic tumult.

A spirit of resilience and learning are emerging in some islands, sparking curiosity in their people to assess whether they can bounce back

---

[11] Visit: https://www.un.org/development/desa/dpad/publication/un-desa-policy-brief-64-the-covid-19-pandemic-puts-small-island-developing-economies-in-dire-straits/.
[12] Visit: https://unctad.org/news/impact-covid-19-tourism-small-island-developing-states.
[13] Visit: https://www.theolivepress.es/spain-news/2020/05/04/balearic-islands-predict-economy-will-shrink-three-times-more-than-rest-of-spain-due-to-impact-of-coronavirus-on-tourism-sector/.
[14] See, for example, Grima, Gonzi, and Thalassinos (2020).
[15] Interestingly, the first country to establish a travel ban worldwide was the Marshall Islands. See Kenny (2021).

and forward, and act as agents of change. Because of their size, initiatives usually reveal the strong public and private partnership supporting fiscal stimulus packages. Most islands engaged in fiscal stimulus to keep the economy going, protecting jobs and businesses, securing household incomes, and coming together for the common good. The pandemic also created the opportunity to accelerate the provision of widespread high-speed internet access to nimble itself to remote education and improve business processes. Small islands in general are laggards in the field of broadband connectivity and outdated information and communications technology infrastructure. However, critical opportunities exist to support the digital transformation, increase economic resilience, and improve responsiveness to natural disasters, to enhance e-learning modalities and to accelerate small business development innovation.[16]

Moreover, some islands invented innovative activities and services such as Barbados Welcome Stamp, which allowed visitors to stay on the island without a visa for up to one year. St. Lucia entered a partnership with the World Economic Forum (WEF) charting a new financing model on its country financing roadmap. In Fiji, more than 170,000 residents joined the "Barter for Better Fiji" Facebook group to develop a transaction economy based on barters. In a webinar about small communities in the Caribbean and their endeavors to overcome the adverse cascading effects of COVID, held in August, 2020, I learned about some innovative initiatives in Jamaica such as the Village Tourism program, the digital diaspora services, and creative ideas of how authentic community stories can blend with crafts. In other webinars, I also learned how small islands in the Caribbean were able to reconfigure their health care sector to prepare and adjust for COVID first-case scenarios. Other innovative ideas were digital platforms to support the most vulnerable groups in small islands, such as the Happy to Give Back Program.

The private sector seems to review what is working and what is not working and based on that review is coming up with a new way of doing things. Consider the Aruba case again. With the advent of the COVID 19 and the financial crisis that followed, most businesses took a hard hit as demand for the products and services dropped below breakeven levels. Amidst this downward trend, I noted several businesses that not only survived this fluid economic environment, but some that are even thriving.

---

[16] See, for example, Giraldo (2018).

While no conclusive evidence exists from my conversations with multiple stakeholders in July and August 2020, I noticed several characteristics that may have contributed to this thriving success.

First and foremost was establishing a flexible pricing schedule that allowed companies to retain fleeting customers that were tightening their budgets. Flexible pricing involves breaking down the product or services into different tiers/levels that would allow the customer to choose what she/he wanted to include or exclude depending on their budget constraints. For example, a security company that would typically offer a single package and price that included an anti-theft, fire, and surveillance system would now split the services into three different parts with the anti-theft system and corresponding price-based service. All other services would be considered add-ons that would increase the price. This pricing mechanism gives the sales representative a powerful tool to negotiate with clients, ultimately resulting in customer retention. As a side-note, the flexible pricing scheme also allows a company to promote its products/services at a lower price than competitors, allowing them to launch promotions even during a crisis.

A second important characteristic of those companies was achieving a very lean operational overhead. These companies outsourced manpower either through flex workforce companies or independent contractors working on a project basis and reduced operational expenses. One example that caught my attention was a company that installed solar panels. Its overhead consisted of only two employees, while the necessary work-force to cater to both commercial and residential clients was outsourced to contractors. This feature gave the company the ability to survive steep decreases in sales.[17] Such resilient recovery activities are essential to face COVID as a transformational event giving way to newer realities in supply and demand dynamics.

8.4   PRIMING THE FUTURE

The recovery will shape tourism demand and supply. Tourism demand and supply characteristics are unique in comparison to other economic sectors. The impact of COVID suggests that the major challenge for small islands

---

[17] My conversation with Dr. Eduardo Parra Lopez also revealed interesting firms' innovative behavior in the Canary Islands.

is an issue of revenues, a lack of demand. However, as soon as the source countries recover from the pandemic, demand will be back, triggering a steady recovery of the tourism industry in small island destinations. Yet, I expect that things will change and business as usual may not be possible. Tourists' preferences may have changed due to the pandemic. Moreover, the economics literature teaches us that behavior is malleable and subject to change. If this is the case, then it is a dangerous proposition to contemplate tourism demand elasticity as a constant. Consider, for example, the relationship between income and human well-being. For the longest time, higher-income was presumed to trigger greater happiness and life satisfaction levels because the belief was that the maximization of growth and consumption were the most vital determinants of well-being. By now, we know that beyond a certain point of prosperity, increases in private consumption or material possessions have little impact on well-being; more income neither makes people happier nor more satisfied with life.

This central finding from a large body of literature is a critical consideration in TS because it pivots vacationing from an everyday life condition to a potential crucial life condition of well-being. This shift from banality to an invigorating life condition augurs well for TS because this shift means that demand elasticity should remain stable. People need to continue to vacation for their mental health and well-being, either to escape, enjoy the beach, re-charge, temporarily change lifestyle, and engage themselves in the nuances of new cultures. Travel is an effective and powerful way to experience differences of contrast to expand our life perspectives. However, the burgeoning tourism literature is showing that demand elasticity is not stable but is changing. This literature argues that business cycles influence this reality, as indicated previously, where the economic ebb and flow affect consumers' behavior. This behavior reveals transitory or permanent patterns, depending on how people make choices or how they are coordinated.[18]

An alternative perspective may involve a natural progression in international tourists' preferences. Vacationing may have shifted from a luxury to a necessity good, implying that residents from rich countries need to

---

[18] Incentives, the impulsiveness of animal spirits, as suggested by Romer's works, and Ackerloff and Shiller, may be the culprits of erratic behavior and may impact the intensity, direction, and magnitude of tourism demand and supply dynamics and economic activities in general. Micro choices, such as loss aversion, liquidity constraints, or habit modifications, influence demand elasticities. More macro forces, including disruptive technology, shape productivity to the benefit or detriment of tourists and residents' welfare.

travel yearlong. The preference is no longer just vacation-vacation appears to have become a natural or human right. Instead, preference refers to how long they will stay, where they want to go, and what product they want to consume. When income in rich countries exceeds a certain income threshold, the quality of the experience becomes more important than how many times a person goes on vacation (quantity). Quality reveals a distinctive property or attribute, which in the case of a tourist destination, turns into memorable experiences, resulting in customer satisfaction, superior value, and competitive advantage. A destination must produce various goods and services with the same quality consistency and do so consistently.

The destination develops its brand-the longevity of which is determined by way of its consistent quality of goods and services. The brand becomes dependent upon its linkages with the small business entrepreneur. This linkage, however, is both frail and intricate. Its frailty arrives in its tenuous dependency on the benefits that loyal tourists will return to purchase their goods and services. Without the assurance of those benefits, the small business may not produce the quality product that the tourist requires. In short, the destination brand does have some dependency on that small business linkage. Thus, for profits' sake, strategic coordination between small businesses and the industry must be groomed and protected. This coordination is a complicated endeavor for both the industry and local business existence.

When analyzing TS, tourism literature usually examined the relationship between tourism development and economic growth. Tourism development builds resources through investment and capital formation and impacts the allocation of available resources. The impact of tourism development may be positive (e.g., jobs and income) or negative (e.g., pollution and crime), influencing the tourist's experience and residents' well-being simultaneously. Tourism development refers to infrastructure, physical, and human assets, or services at a destination that promotes tourist arrivals and spending. From this perspective, tourism development is a multidimensional construct, including economic, social, cultural, and environmental conditions. Arguably, TS has an impact on tourism development.

The TS literature stream's central premise is that the positive growth performance could be a sustained achievement if small countries can import growth from abroad through continued gains in ToT. This premise is only an opportunity and possibility when growth in small islands results from a continued appreciation of tourism (experience and services) rather than physical expansion and more arrivals. There is a clear consensus regarding the dynamic endogenous force of TS, provoking growth in the short-term. The long-term effects of TS, however, are tenuous.[19]

The question is whether the positive trend can continue forever or whether the trend patterns might reach a threshold, after which the threshold will bend downward. In other words, it is not clear if TS reveals a sustained upward trend or only a transitory upward trend. Some studies examining this TS relationship with economic growth argue that a positive relationship is not forever, but it will experience a negative trend over time. In other words, TS may favor economic growth in small countries at the beginning. However, this activity may have a similar effect on the economy than other industries when a certain development level is achieved.[20]

8.5   CREATING A LEARNING DESTINATION

Of course, if comparative advantage is a dynamic process, a destination can move from a low-performing comparative advantage to a higher-performing comparative advantage. This movement assigns time as a critical component in the TS process, suggesting that the TS process could entertain cyclical (rise and fall) patterns. The dynamic process stems from the acquisition of a future advantage ensconced in higher productivity growth. This process change is ongoing because it results from businesses, government, and individual actors' everyday deliberate choices. These choices may include routine changes such as changing contracts, organizing a new event, or focusing on a new segment. However, the most fundamental long run effects of change come from the learning of these actors. Therefore, the demand level is endogenously determined by the residents' physical and human capital accumulation choices. This choice cannot be determined a priori. Its effects depend on what and how residents learn from this specialization experience and how learning can be put into practice to grow and advance human development.[21]

---

[19] For a discussion on the dynamic effects of tourism specialization, see, for example, Sequeira and Campos (2005); Brau et al. (2007); Schubert et al. (2011); and Marsiglio (2018).
[20] See, for example, Figini and Vici (2010).
[21] See, for example, Stiglitz and Greenwald (2014).

One example of learning in tourism is the illustration expounded by Lejárraga and Walkenhorst. These researchers posit that small island business owners quickly learn the wants and needs of their tourists. It follows then that these entrepreneurs must learn and learn fast the need of the tourism venue to keep pace. Thus, they must respond with smart, cost-effective decisions in response to tourists' preferences. The likelihood of positive economic growth is then set in motion. If the response requires entrepreneurs to access other local businesses such as catering, travel accommodations, event offerings, and safety considerations, then these interconnections can further contribute to economic growth. The capacity of the tourism sector depends on the fragmentation level of its production process. Competition arises between industry sectors-sectors necessary for the tourism business's overall configuration, such as hotel accommodations, an airlift to and from the destination, restaurants, shopping, and other amenities and infrastructure are necessary on-site tourist sectors. Making available extremely diverse factors according to tourist choice could lead to an inability to coordinate and control the type of superior product that makes up a memorable experience marketed to the tourist.

The destination may contain a comparative advantage in one or two segments of the production process while having a disadvantage in another. For example, a destination may have a comparative advantage in its attraction and entertainment but may reveal a lack of advantage in the accommodation sector. Nowak, Petit, and Sahli show how destinations can move up or down the production process ladder or move in the opposite direction of the specialization scale and scope.[22] The comparative advantage of a destination, thus, can change over time. However, the time constraint embedded in the simultaneous production and consumption of the tourism product (experience) means that tourism may be entirely determined by demand. This reasoning stems from the nature of the experience as a perishable product. Perishability determines that the experience, in essence, is a short-term product. Thus, time, and learning play a significant role in determining the change process.

However, the process is also influenced by endogenous aspects embedded in the tourism product. Demand studies mainly view the tourism product as a result of exogenous factors. From this perspective, tourist arrivals

---

[22] See Nowak, Petit, and Sahli (2010).

seem to obey exogenous factors (e.g., the macroeconomic conditions in source countries) that lie beyond destinations' control. Arguably, tourist arrivals also depend on quality indicators at the destination. The destination endogenously determines these quality indicators. Destination quality indicators include facilities, quality of hotels and services, environmental quality, cultural heritage, and political stability and capabilities. These elements have an endogenous dimension because they depend on vision, commitment, and investment.[23] The arrival numbers are thus endogenously determined by the residents' choices regarding physical and human capital accumulation.[24] Quality also has another important dimension in the production process. It may suppress the incentive of tourist suppliers to cheat on the quality of the product. This cheating incentive stems from the transient nature of the tourism product.[25]

Small island destinations cannot depend on tourist visits as a means to secure sustained economic growth. Over time, policies dictating supply must assume greater realms of responsibility. Supply policies and constraints could dictate less appeal to the consumer if left to stagnate into a been-there-done-that scenario. Instead, small island destinations would need to find the means to address tourist preferences with experiences comprised of unique goods and services that are marketable and desirable internationally. Given their natural resources of sun, sand, and sea and the utopian flavor of paradise, the small island has attracted a tourism industry market. However, despite its magnetism, the small island destination cannot bear up under continually increasing tourist visits. Its size, as well as its supply constraints, prohibit that option. Strategic prospects must be developed in light of this. Mauritius developed its strategy by shifting its focus from attracting big tourist numbers to developing a more fastidious approach to tourist acquisition. It converted its market from high tourist volume and tourist spending ability to a larger, more varied market segment. The result allowed Mauritius to reduce leakage and continue to drive its economic growth and suggest that international tourist arrivals and residents' income not necessarily move in the same direction. Shifting from volume to quality presupposes an understanding

---

[23] Several studies have emphasized this endogenous dimension of tourism production. See, for example, Alegre and Pou (2006); and Eugenio-Martin et al. (2008).

[24] This assertion follows Lucas's endogenous model's logic. See Lucas (1988).

[25] For example, Keane (1997) discusses this moral hazard problem of tourism and its potential adverse effect on the tourist experience's quality assurance.

that the destination should pivot to income growth at a specific arrivals level.[26]

As discussed in the previous chapter, TS involves threshold patterns. Because specialization (mentioned in Chapter 2) is a collective process, pivoting requires coordination acumen, and this a hard and sensitive task. It is finding out what must be done, who should be involved (which stakeholders), reaching an agreement on how to undertake the activities, and who benefits and loses. Again, there is only one way to perfection and a million ways to failure (remember Tolstoy in Anna Karenina!). Successfully navigating this response to a competitive market that increasingly requires differentiated goods and services that engender greater tourist loyalty to a destination could balance the increase in the cost of producing such goods and services. To meet the aggregate of such a task requires the whole of the human spirit to conceive the means to emerge with successful research and development strategies that go forward. This creativity hinges on a continuous focus on innovation (the creation and sustainability of an innovative ecosystem) and inclusiveness. The innovation articulation should be geared to expand the variety of available goods (the product sophistication), which depends on collective efforts. Trade between countries is motivated by a preference for diversity.

8.6 THE BIG TAKE AWAY

My personal experience and research have taught me that small islands must continuously prove their viability; otherwise, they could be ignored entirely. TS is a roadmap towards this viability. If its dynamics are well understood, are strategically planned, and acted upon, it can provide sustained economic growth and provide great utility externality for residents and tourists alike. The roadmap consists of two interdependent components: a new business model and a measure of what counts. The new business model centers on increasing terms of trade (ToT), which consists of scarcity and product sophistication, and a demand elasticity based on the concept of luxury. The primary measurement of a sustained model is to gauge the ToT continuously; as long as the ToT is positive,

[26] Marsiglio (2018) provides the theoretical foundations undergirding the claim that arrivals and residents' income may move in the opposite direction without harming the small island.

small islands can scale up and push economic growth and prosperity. TS has worked in the past and will work in the future.

The business model that stems from my analysis throughout this book uses volume through TS to launch development during the switch to a scarcity and sophistication model. This model has the seeds for a steady pathway for a diversified economy because sophistication requires more quality semi-products and services, and hence more backward and forward economic linkages. The application of scarcity should come naturally to a small island. Geography ultimately will impose its whims: a small island can only build so many hotel rooms; it will run out of space. Alternatively, sophistication is a matter of free choice: it depends on investments, resources, efforts, and optimal time use. Sophistication also means the choice of learning how to cope, mitigate, and manage risks. The sudden reversal of tourism flows can be reduced and reversed through the development of appropriate strategies.

TS is a means to an end: to scale up and to enhance human development. TS cannot solve the small island vulnerabilities; it provides opportunities, possibilities, and choices that small islands can take to face the future. I know from experience and through research that small islands tout their people as the core and scope of their development. People refer to talents, skills, dedication, and wisdom to foster coordination processes to solve collective problems supporting resilience. The investments in time and efforts to improve coordination efforts enhance trust and institutional strength. Coordination requires balancing self-interest with collaboration and involves an enormous capacity to establish trust, which is the hallmark for broadening social and economic exchanges. Without trust, businesses cannot plan effectively because they do not know who and what can be trusted; and lack of trust increases transaction costs.

I am optimistic about the future of small islands because of their intangible wealth. They are experienced in the tourism business. They understand and know that the tourism product requires everybody to work together: the value of their input depends on others' input. They understand that this coming together, if not done smoothly, can create a gridlock; therefore, learning coordination skills is vital to be successful.[27] So, by nature and by

---

[27] Candela and Figini (2010) define the tourism product as an anti-common, which means the multitude of suppliers need to agree or complement each other to generate the product. Without coordination, the tourist product is not possible. Heller (2008) characterizes a fragmented economic system as a free market paradox. According to him, "when too many people own pieces of one thing, cooperation breaks down, wealth disappears, and everybody loses" (p. xiv).

choice small islands should practice scarcity. They should work together to scale up and open themselves to the world. TS is a pathway to scale up, to grow; and the tourism product by its very nature requires a delicate balance between self-interest and collaboration. Small islands are up to this delicate task because they know how to make a tourist product. They learned through trying and experience that, when dealing with a perishable goods such as tourism, tourists will use reputation mechanisms to determine which destination to visit. One of those mechanisms is the digital eco-system. The Corona pandemic has shown the critical need to accelerate the system in order to support health services needing to provide real time information regarding tourist activities, services, operations, risks, etc. Moreover, reputation is the reciprocal coexistence of many small firms that interact and coordinate with each other based on trust to generate and deliver the quality tourism product.

The reputation of a small island destination and its constituting firms is interdependent meaning that the destination depends on how firms behave towards the tourists. Small island firms understand that they are too small to promote their reputation: it is simply too costly. So, they depend on the reputation of the destination. They have an incentive to protect the destination's reputation because a collective bad reputation jeopardizes business opportunities with others due to a lack of trust and a negative stereotype. Reputation thus is a public good.[28] As long as the small island can create an environment where it can innovate to enhance this collective reputation, it will prosper. The recipe is to marry individual behavior with collective behavior. Moreover, the pandemic unleashes the opportunity to break away from old habits and biases that have favored the status quo.

There is a pathway to the future for small islands. Honing in on TS is a choice that small islands could make to scale up, grow, and instigate human development and welfare. TS can spark a concentric circle of knowing what we can do best in accordance with that which is best suited to our conditions. It is learning from our experiences that leads us and sanctions what we choose to do. It is a change in mindset and practice that frees us from the roots of learning to survive. Rather, it thrusts us into the visages of collaborative innovation and the diffusion and sharing of the ideas that propels us from cowed beginnings to the realization that we can forge a path to our own prosperity.

---

[28] Tirole (2017) refers to reputation as a common good.

KEYWORDS

- Canary Islands
- gross domestic product
- learning destination
- potential COVID Impact
- tourism specialization
- world tourism organization

BIBLIOGRAPHY

Acevedo, S., (2014). *Debt, Growth and Natural Disasters: A Caribbean Trilogy*. IMF Working Paper, No. 14/125, Washington, DC: IMF.

Adamou, A., & Clerides, S., (2010). Prospects and limits of tourism-led growth: The international evidence. *Review of Economic Analysis, 2*(3), 287–303.

Adams, P., & Parmenter, B., (1992). *The Medium-Term Significance of International Tourism for the Australian Economy*. Canberra: Bureau of Tourism Research.

Aguiló, E., Alegre, J., & Sard, M., (2005). The persistence of the sun and sand tourism model. *Tourism Management, 26*(2), 219–231.

Alegre, J., & Pou, L., (2006). The length of stay in the demand of tourism. *Tourism Management, 27*, 1343–1355.

Alesina, A., & Spolaore, E., (2003). *The Size of Nations*. Cambridge, MA: MIT press.

Alesina, A., (2003). The size of countries: Does it matter? *Journal of the European Economic Association, 1*(2, 3), 301–316.

Alesina, A., Spolaore, E., & Wacziarg, R., (2000). Economic integration and political disintegration. *American Economic Review, 90*, 1276–1296.

Algieri, A., Aquino, A., & Succurro, M., (2018). International competitive advantage in tourism: An eclectic view. *Tourism Management Perspectives, 25*, 41–52.

Algieri, B., (2006). International tourism specialization of small countries. *International Journal of Tourism Research, 8*(1), 1–12.

Álvarez-Albelo, C. D., & Hernández-Martín, R., (2009). Specialization in luxury goods, productivity gaps and the rapid growth of small tourism countries. *Tourism Economics, 15*(3), 567–589.

Apostolopoulos, Y., & Gayle, D., (2002). *Island Tourism and Sustainable Development: Caribbean, Pacific and Mediterranean Experiences*. Westport: Praeger.

Areski, R., Cherif, R., & Piotrowski, J., (2009). *Tourism Specialization and Economic Development*. Washington DC: IMF Working Paper, WP/09/176.

Armstrong, H., & Read, R., (2000). Comparing the economic performance of dependent territories and sovereign micro-states. *Economic Development and Cultural Change, 48*(2), 285–306.

Armstrong, H., & Read, R., (2001). *Globalization and Economic Development: Lessons from Small States*. Paper for conference small states in the world markets-15 years later, Goteborg University.

Armstrong, H., & Read, R., (2003). The determinants of economic growth in small states. *The Round Table, 368*, 99–124.

Armstrong, H., De Kervenoael, R., Li, X., & Read, R., (1998). A comparison of the economic performance of different microstates and between microstates and larger countries. *World Development, 26*(4), 639–656.

Arrow, K., (1962). The economic implications of learning by doing. *Review of Economic Studies, 29*, 155–173.

Ashoff, G., (1989). *Economic and Industrial Development Options for Small Third World Countries.* Occasional Paper No. 91. Berlin: German Development Institute.

Balaguer, J., & Cantavella-Jorda, M., (2002). Tourism as a long-run economic growth factor: The Spanish case. *Applied Economics, 34*(7), 877–884.

Balassa, B., (1965). Trade liberalization and revealed comparative advantage. *The Manchester School, 33*(2), 99–123.

Baldacchino, G., & Milne, D., (2000). *Lessons from the Political Economy of Small Islands. The Resourcefulness of Jurisdiction.* London: McMillan Press Ltd.

Baldacchino, G., (2004). The coming of age of island studies. *Tijdschrift Voor Economische en Sociale Geografie, 95*(3), 272–284.

Baldacchino, G., (2006). Managing the hinterland beyond: Two ideal-type strategies of economic development for small island territories. *Asia Pacific Viewpoint, 47*(1), 45–60.

Baldacchino, G., (2006). Warm versus cold water island tourism: A review of policy implications. *Island Studies Journal, 1*(2), 183–200.

Baldacchino, G., (2007). Islands as novelty sites. *Geographical Review, 97*(2), 165–174.

Banerjee, A., & Duflo, E., (2019). *Good Economics for Hard Times.* New York: Public Affairs.

Barfield, S., (2003). *Development, the World Trade Organization and the 'Banana Trade War.'* Unpublished; PhD dissertation, Sheffield: University of Sheffield.

Barro, R., (1991). Economic growth in a cross section of countries. *Quarterly Journal of Economics, 106*(2), 407–443.

Benedict, B., (1967). Sociological characteristics of smaller territories and their implications for economic development. In: Banton, M., (ed.), *The Social Anthropology of Complex Societies* (pp. 23–36). London, UK: Tavistock.

Bernal, R., Bryan, T., & Fauriol, G., (2001). *The United States and Caribbean Strategies: Three Assessments.* Policy Papers on the Americas. Washington DC: CSIS Americas Program.

Bertram, G., & Poirine, B., (2007). Island political economy. In: Baldacchino, G., (ed.), *A World of Islands: An Island Studies Reader* (pp. 325–378). Malta: Institute of Island Studies and Agenda Academic.

Bertram, G., (1986). Sustainable development in Pacific micro-economies. *World Development, 14*(7), 889–992.

Bertram, G., (2004). On the convergence of small island economies with their metropolitan patrons. *World Development, 32*(2), 343–364.

Bhagwati, J., & Srinivasan, T., (1979). Trade policy and development. In: Dornbusch, R., & Frenkel, J., (eds.), *International Economic Policy: Theory and Evidence* (pp. 1–35). Baltimore: Johns Hopkins University Press.

Bishop, M., (2010). Tourism as a small-state development strategy: Pier pressure in the Eastern Caribbean? *Progress in Development Studies, 10*(2), 99–114.

Boissevain, J., & Theuma, N., (1998). Contested space, planners, tourists, developers and environmentalists in Malta. In: Abram, S., & Waldren, J., (eds.), *Anthropological Perspectives on Local Development* (pp. 96–119). London: Routledge.

Bramwell, B., (2003). Maltese responses to tourism. *Annals of Tourism Research, 30*(3), 581–605.

Brau, R., Lanza, A., & Pigliaru, F., (2003). *How Fast are the Tourism Countries Growing? The Cross-Country Evidence.* Working Papers 2003.85, Fondazione Eni Enrico Mattei.

Brau, R., Lanza, A., & Pigliaru, F., (2005). An investigation on the growth performance of small tourism countries. In: Lanza, A., Markandya, A., & Pigliaru, F., (eds.), *The*

Economics of Tourism and Sustainable Development (pp. 8–29). Cheltenham, UK: Edward Elgar.

Brau, R., Lanza, A., & Pigliaru, F., (2007). How fast are small tourism countries growing? Evidence from the data for 1980–2003. *Tourism Economics, 13*(4), 603–613.

Brida, J., & Pulina, M., (2010). *A Literature Review on the Tourism-Led-Growth Hypothesis.* CRENOS Working Papers, Cagliari, Italy.

Briguglio, L., (1995). Small island developing states and their economic vulnerabilities. *World Development, 23*(9), 1615–1632.

Briguglio, L., (1998). Small country size and returns to scale in manufacturing. *World Development, 26*(3), 507–515.

Briguglio, L., Persaud, B., & Sters, R., (2005). *Toward an Outward-Oriented Development Strategy for Small States: Issues, Opportunities, and Resilience.* World Bank/International Monetary Fund.

Britton, S., (1982). The political economy of tourism in the third world. *Annals of Tourism Research, 9*(3), 311–358.

Brown, D., (2010). Institutional development in small states: Evidence from the commonwealth Caribbean. *Halduskultuur- Administrative Culture, 11*(1), 44–65.

Brundenius, C., (2002). *Tourism as an Engine of Growth: Reflections on Cuba's New Development Strategy.* Copenhagen, CDR Working Paper 02.10.

Bryan, A., (2001). *Caribbean Tourism: Igniting the Engines of Sustainable Growth.* Miami: Miami University.

Bryden, J., (1973). *Tourism and Development: A Case Study of the Commonwealth Caribbean.* Cambridge UK: Cambridge University Press.

Buswell, R., (2011). *Mallorca and Tourism, History, Economy and the Environment.* Bristol UK: Channel View Publications.

Butcher, J., (2003). *The Moralization of Tourism: Sun, Sand and Saving the World.* London: Routledge.

Butler, R., (1993). Tourism development in small islands: Past influences and future directions. In: Lockhart, D., Drakakis-Smith, D., & Schembri, J., (eds.), *The Development Process in Small Island States* (pp. 71–91). London: Routledge.

Candela, G., & Cellini, R., (2006). Investment in tourism market: A dynamic model of differentiated oligopoly. *Environmental and Resource Economics, 35*(1), 41–58.

Candela, G., & Figini, P., (2010). *The Economics of Tourism Destinations.* Dordrecht, Netherlands, Springer.

Capó, J., Font, A., & Nadal, J., (2007). Dutch disease in tourism economies: Evidence from the Balearics and the Canary Islands. *Journal of Sustainable Tourism, 15*(6), 615–627.

Carey, K., (1991). Estimation of the Caribbean tourism demand: Issues in measurement and methodology. *Atlantic Economic Journal, 19*(3), 32–40.

Ceata-Hatton, M., (1997). *Towards a Sustainable Tourism Zone in the Wider Caribbean.* Sto. Domingo: Cieca.

Cerina, F., (2007). Tourism specialization and environmental sustainability in a dynamic economy. *Tourism Economics, 13*(4), 553–582.

Clancy, M., (1998). Commodity chains, services and development: Theory and preliminary evidence from the tourism industry. *Review of International Political Economy, 5*(I), 122–148.

Clancy, M., (1999). Tourism and development, evidence from Mexico. *Annals of Tourism Research, 26*(1), 1–20.

Clancy, M., (2001). *Exporting Paradise: Tourism and Development in Mexico.* Oxford, UK: Elsevier.

Clarke, C., (1978). *An Analysis of the Determinants of Demand for Tourism in Barbados.* Unpublished Ph.D dissertation, Fordham University.

Clarke, C., Wood, C., & Worrell, D., (1986). Prices, income and the growth of tourism in Barbados. *Economic Review, XIII*(I), 0–45.

Cole, S., & Razak, V., (2009). How far, and how fast? Population, culture and carrying capacity in Aruba. *Futures, 41*, 414–425.

Cole, S., (2007). Beyond the resort life cycle: The micro-dynamics of destination tourism. *The Journal of Regional Analysis & Policy, 37*(3), 266–278.

Conlin, M., & Baum, T., (1995). *Island Tourism: Management, Principles and Practice.* Sussex: John Wiley and Sons.

Connell, J., & Conway, D., (2000). Migration and remittances in island microstates: A comparative perspective on the South Pacific and the Caribbean. *International Journal of Urban and Regional Research, 24*(1), 52–78.

Copeland, B., (1991). Tourism, welfare and de-industrialization in a small open economy. *Economica, 58*, 515–529.

Croes, M., (2011). *De Weg Naaar the Status Aparte: De Laatste Etappe.* Nijmegen, the Netherlands: Wolf Legal Publishers.

Croes, R., & Moenir, A. L., (1990). De-colonization of Aruba within the Netherlands Antilles. In: Sedoc-Dahlberg, B., (ed.), *The Dutch Caribbean: Prospects for Democracy.* New York, NY: Gordon and Breach.

Croes, R., & Ridderstaat, J., (2017). The effects of business cycles on tourism demand flows in small island destinations. *Tourism Economics, 23*(7), 1451–1475.

Croes, R., & Ridderstaat, J., (2018). Tourist motivation and demand for islands. In: McLeod, M., & Croes, R., (eds.), *Tourism Management in Warm-water Island Destinations.* CABI: Wallingford, UK.

Croes, R., & Schmidt, P., (2007). Promoting tourism as US foreign aid: Building on the promise of the Caribbean basin initiative. *Journal of Multidisciplinary Research, 1*(1), 1–15.

Croes, R., & Tesone, D., (2007). The indexed minimum wage and hotel compensation strategies. *Journal of Human Resources in Hospitality and Tourism, 6*(1), 109–124.

Croes, R., & Vanegas, M., (2000). Evaluation of demand, US Tourists to Aruba. *Annals of Tourism Research, 27*(4), 946–963.

Croes, R., & Vanegas, M., (2003). Growth, development and tourism in a small economy: Evidence from Aruba. *International Journal of Tourism Research, 5*(5), 315–330.

Croes, R., & Vanegas, M., (2005). An econometric study of tourist arrivals in Aruba and its implications. *Tourism Management, 26*(6), 879–890.

Croes, R., & Vanegas, M., (2008). Cointegration and causality between tourism and poverty reduction. *Journal of Travel Research, 47*(1), 94–103.

Croes, R., (2005). Value as a measure of tourism performance in the era of globalization: Conceptual considerations and empirical findings. *Tourism Analysis, 9*(4), 255–267.

Croes, R., (2006). A paradigm shift to a new strategy for small island economies: Embracing demand side economics for value enhancement and long-term economic stability. *Tourism Management, 27*(3), 453–465.

Croes, R., (2010). *Anatomy of Demand in International Tourism: The Case of Aruba.* Saarbrucken, Germany: Lambert Academic Publishing.

Croes, R., (2011). Measuring and explaining competitiveness in the context of small island destinations. *Journal of Travel Research, 50*(4), 431–442.

Croes, R., (2011). *The Small Island Paradox: Tourism Specialization as a Potential Solution.* Saarbrucken, Germany: Saarbrucken, Germany: Lambert Academic Publishing.

Croes, R., (2012). Assessing tourism development from Sen's capability approach. *Journal of Travel Research, 51*(5), 542–554.

Croes, R., (2013). Tourism specialization and economic output in small island destinations. *Tourism Review, 68*(4), 34–48.

Croes, R., Ridderstaat, R., & Van, N. M., (2017). Connecting quality of life, tourism specialization, and economic growth in small island destinations: The case of Malta. *Tourism Management, 65,* 212–223.

Croes, R., Rivera, M., & Semrad, K., (2011). *Winning the Future.* Orlando, FL, Dick Pope Sr. for Tourism Studies.

Croes, R., Rivera, M., Semrad, K., & Khalilzadeh, J., (2017). *Happiness and Tourism in Aruba: Insights from the 2016 Happiness Survey.* Orlando, FL: Dick Pope Sr. Institute for Tourism Studies.

Crowards, T., (2002). Defining the category of 'small' states. *Journal of International Development, 14*(2), 143–179.

Cuadrado-Roura, J., & Rubalcaba-Bermejo, L., (1998). Specialization and competition amongst European cities: A new approach through fair and exhibition activities. *Regional Studies, 32*(2), 133–147.

De Kadt, E., (1979). *Tourism: Passport to Development?* Oxford: Oxford University Press.

Demas, W., (1965). *The Economics of Development in Small Countries with Special Reference to the Caribbean.* Montreal: McGill University Press.

Demas, W., (1992). The post-independence Caribbean: Development and survival. *Caribbean Affairs, 5*(3), 1–11.

Dogru, T., Sirikaya-Turk, E., & Crouch, G., (2017). Remodeling international tourism demand: Old theory and new evidence. *Tourism Management, 60,* 47–55.

Dreher, A., (2006). Does globalization affect growth? Evidence from a new index of globalization. *Applied Economics, 38*(10), 1091–1110.

Dritsakis, N., (2004). Tourism as a long-run economic growth factor: An empirical investigation for Greece using causality analysis. *Tourism Economics, 10*(3), 305–316.

Durbarry, R., (2004). Tourism and economic growth: The case of Mauritius. *Tourism Economics, 10*(4), 389–401.

Dutta, S., Bergen, M., Levy, D., Ritson, M., & Zbaracki, M., (2002). Pricing as a strategic capability. *MIT Sloan Management Review,* 61–66.

Dwyer, L., & Forsyth, P., (1998). Estimating the employment impacts of tourism to a nation. *Tourism Recreation Research, 23*(2), 1–12.

Dwyer, L., Forsyth, P., Madden, J., & Spurr, R., (2000). Economic impacts of inbound tourism under different assumptions regarding the macroeconomy. *Current Issues in Tourism, 3*(4), 325–363.

Easterly, W., & Kraay, A., (2000). Small states, small problems? Income, growth and volatility in small states. *World Development, 28*(11), 2013–2027.

Easterly, W., & Kraay, A., (2001). *Growth is Good for the Poor.* World Bank Policy Research, Working Paper 2587, Washington DC: The World Bank.

Easterly, W., & Levine, R., (2003). Tropics, germs, and crops: How endowments influence economic development. *Journal of Monetary Economics, 50*(1), 3–39.

Easterly, W., (2001). *The Elusive Quest for Growth: Economists' Adventures and Misadventures in the Tropics.* Cambridge, Mass: MIT Press.

Escaith, H., (2001). Las economias pequenas de America Latina y el Caribe. *Revista CEPAL, 74,* 71–85.

Eugenio-Martin, J., Martin-Morales, N., & Scarpa, R., (2004). *Tourism and Economic Growth in Latin American Countries: A Panel Data Approach.* FEEM Working Paper No. 26, Milan.

Eugenio-Martin, J., Martin-Morales, N., & Sinclair, M., (2008). The role of economic development in tourism demand. *Tourism Economics, 14*(4), 673–690.

Everest-Phillips, M., (2014). *Small, so Simple? Complexity in Small Island Developing States.* UNDP. https://www.undp.org/content/dam/undp/library/capacity-development/English/Singapore%20Centre/GPCSE_Complexity%20in%20Small%20Island.pdf (accessed on 4 October 2021).

Fagence, M., (1999). Tourism as a feasible option for sustainable development in small island developing states (SIDS): Nauru as a case study. *Pacific Tourism Review, 3*(2),133–142.

Fagerberg, J., (2003). *Innovation: A Guide to the Literature.* https://smartech.gatech.edu/bitstream/handle/1853/43180/JanFagerberg_1.pdf?sequence=1&isAllowed=y (accessed on 4 October 2021).

Figini, P., & Vici, L., (2010). Tourism and growth in a cross-section of countries. *Tourism Economics, 16,* 789–805.

Fossati, A., & Panella, G. (2000). *Tourism and Sustainable Economic Development* (pp. 57–69). Dordrecht, Neth.: Kluwer.

Frankel, J., & Romer, D., (1999). Does trade cause growth? *American Economic Review, 89*(3), 379–399.

Freckleton, M., (2000). Liberalization of trade in services and diversification of Caricom exports. *Global Development Studies, 2,* 108–123.

Gal, M., (2003). *Competition Policy for Small Market Economies.* Cambridge, Massachusetts: Harvard University Press.

Gatt, E., (20014). *The Human Development Index and Small States.* Occasional Papers on Islands and Small Islands, No:5/2004, ISSN 1024-6282, University of Malta.

George, A. L., (2019). Case studies and theory development: The method of structured, focused comparison. In: Caldwell, D., (ed.), *Alexander L. George: A Pioneer in Political and Social Sciences. Pioneers in Arts, Humanities, Science, Engineering, Practice* (Vol 15). Springer, Cham.

Gillis, J., (2007). Island sojourns. *Geographical Review, 97*(2), 274–287.

Giraldo, C., (2018). *Could Digital Transformation Help the Caribbean Become more Resilient to Natural Disasters?* Inter-American Development Bank, Washington, DC, https://blogs.iadb.org/caribbean-dev-trends/en/could-digital-transformation-help-the-caribbean-become-more-resilient-to-natural-disasters/ (accessed on 4 October 2021).

Goede, M., (2016). The history of public administration in the Dutch Caribbean. In: Minto-Coy, I., & Berman, E., (eds.), *Public Administration and Policy in the Caribbean* (pp. 77–94). London: CRC Press,.

Goldstone, P., (2001). *Making the World Safe for Tourism.* New Haven: Yale University Press.

Gossling, S., (2003). Tourism and development in tropical islands: Political ecology perspectives. In: Gossling, S., (ed.), *Tourism and Development in Tropical Islands: Political Ecology Perspectives.* Cheltenham, UK: Edward Elgar.

Gouveia, P., & Rodrigues, P., (2005). Dating and synchronizing tourism growth cycles. *Tourism Economics, 11*(4), 501–515.

Grassl, W., (2003). Why and how to specialize in tourism: An agenda for Jamaica. In: McDavid, H., (ed.), *The Role of Government in Tourism.* Kingston: University of the West Indies Press.

Greenidge, K., (2001). Forecasting tourism demand: An STM approach. *Annals of Tourism Research, 23*(4), 739–754.

Griffith, W., (2002). A tale of four Caricom countries. *Journal of Economic Issues, 36*(1), 79–106.

Grima, S., Gonzi, R., & Thalassinos, E., (2020). *The Impact of COVID-19 on Malta and its Economy and Sustainable Strategies.* https://papers.ssrn.com/sol3/papers.cfm?abstract_id=3644833 (accessed on 4 October 2021).

Guizzardi, A., & Mazzocchi, M., (2010). Tourism demand for Italy and the business cycle. *Tourism Management, 31*(3), 367–377.

Hallak, J., (2010). A product-quality view of the Linder hypothesis. *The Review of Economics and Statistics, 92*(3), 453–466.

Harden, S., (1985). *Small is Dangerous: Microstates in a Macro World.* London: Frances Pinter.

Harrison, A., & Hanson, G., (1999). Who gains from trade reform? Some regaining puzzles. *Journal of Development Economics, 59*(1), 125–154.

Hassan, S., (2000). Determinants of market competitiveness in and environmentally sustainable tourism industry. *Journal of Travel Research, 38,* 239–245.

Hawkins, D., & Mann, S., (2007). The World Bank's role in tourism development. *Annals of Tourism Research, 34*(2), 348–363.

Hazari, B., & Sgro, P., (1995). Tourism and growth in a dynamic model of trade. *Journal of International Trade and Economic Development, 4,* 243–252.

Hein, P., (2004). *Is a Special Treatment of Small Island Developing States Possible?* New York: UNCTAD Publications.

Heller, M., (2008). *Gridlock Economy. How too Much Ownership Wrecks Markets, Stops Innovation, and Costs Lives.* New York: Basic Books.

Helpman, E., (2004). *The Mystery of Economic Growth.* Cambridge, Mass. Belknap Press of Harvard University Press.

Hernández-Martín, R., (2008). Structural change and economic growth in small island tourism countries. *Tourism and Hospitality Planning & Development, 5*(1), 1–12.

Hirschman, A., (1984). Against parsimony: Three easy ways of complicating some categories of economic discourse. *Psychological and Sociological Foundations, 74*(2), 89–96.

Holder, J., (1996). Maintaining competitiveness in a new world order: Regional solutions to Caribbean tourism sustainability problems. In: Harris, L. C., & Husbands, W., (eds.), *International Case Studies in Tourism Planning. Policy and Development* (pp. 145–173). New York: John Wiley.

Holzner, M., (2011). Tourism and economic development: The beach disease? *Tourism Management, 32*(4), 922–933.

Hong, W., (2008). *Competitiveness in the Tourism Sector: A Comprehensive Approach from Economic and Management Points*. Heidelberg, Germany: Physica-Verlag.

Inchausti-Sintes, F., (2015). Tourism: Economic growth, employment and Dutch disease. *Annals of Tourism Research, 54*, 172–189.

Jackman, M., & Lorde, T., (2010). On the relationship between tourist flows and household expenditure in Barbados: A dynamic OLS approach. *Economics Bulletin, 30*(1), 472–481.

Jackman, M., Lorde, T., Lowe, S., & Alleyne, A., (2011). Evaluating tourism competitiveness of small island developing states: A revealed comparative advantage approach. *Anatolia, 22*, 350–360.

Jayawardena, C., & Ramajeesingh, D., (2003). Performance of tourism analysis: A Caribbean perspective. *International Journal of Contemporary Hospitality Management, 15*(3), 176–179.

Jin, J., (2004). On the relationship between openness and growth in China: Evidence from provincial time series. *The World Economy, 27*(10), 1571–1582.

Kaldor, N., (1966). *Causes of the Slow Rate of Economic Growth of the United Kingdom: An Inaugural Lecture.* Cambridge: Cambridge University Press.

Kaldor, N., (1970). The case for regional policies. *Scottish Journal of Political Economy, 17*(3), 337–348.

Kaldor, N., (1985). *Economics Without Equilibrium.* Cardiff UK: University College Cardiff Press.

Kammas, M., (1991). *Export-led Growth in Micro-States: The Case of Cyprus.* PhD Dissertation, the University of Utah.

Keane, M., (1997). Quality and pricing in tourism destinations. *Annals of Tourism Research, 24*(1), 117–130.

Keller, P., & Bieger, T., (2007). *Productivity in Tourism. Fundamentals and Concepts for Achieving Growth and Competitiveness.* Berlin: Erich Schmidt Verlag.

Kenny, C., (2021). *The Plague Cycle. The Unending War between Humanity and Infectious Disease.* New York: Scribner.

Keynes, J., (1932). Economic possibilities for our grandchildren (1962). In: *Essays in Persuasion* (pp. 358–373). New York: Harcourt Brace.

Kim, H., Chen, M., & Jang, S., (2006). Tourism expansion and economic development: The case of Taiwan. *Tourism Management, 27*(5), 925–933.

Krueger, A., (1980). Trade policy as an input to development. *American Economic Review, 70*, 188–292.

Krueger, A., (1985). Import substitution versus export promotion. *Finance and Development, 22*, 20–24.

Kuhn, T., (1962). *The Structure of Scientific Revolutions.* Chicago, IL: University of Chicago Press.

Kuznets, S., (1960). Economic growth in small nations. In: Robinson, E., (ed.), *The Economic Consequences of the Size of Nations* (pp. 14–32). Proceedings of a conference held by the international economic association. London: Macmillan.

Laframboise, M., Mwase, N., Park, M., & Zhou, Y., (2014). *Revisiting Tourism Flows to the Caribbean: What is Driving Arrivals?* IMF Working Paper, WP/14/229, Washington DC.

Lanza, A., (1998). *Tourism Specialization and Economic Growth.* Dissertation, University College of London.

Lanza, A., & Pigliaru, F., (2000). Why are tourism countries small and fast growing? In: Fossati, A., & Panella, G., (eds.), *Tourism and Sustainable Economic Development* (pp. 57–69). Dordrecht, the Neth.: Kluwer.

Lanza, A., Temple, P., & Urga, G., (2003). The implications of tourism specialization in the long run: An economic analysis for 13 OECD economies. *Tourism Management, 24*(3), 315–321.

LaSalle, D., & Britton, T., (2003). *Priceless: Turning Ordinary Products into Extraordinary Experiences.* Boston, Mass: Harvard Business School Press.

Laurent, E., (2014). *Economic Consequences of the Size of Nations, 50 Years on.* Hall 00972823, https://hal-sciencespo.archives-ouvertes.fr/hal-00972823/document (accessed on 4 October 2021).

Lejárraga, I., & Walkenhorst, P., (2010). On linkages and leakages: Measuring the secondary effects of tourism. *Applied Economics Letters, 17*(5), 417–421.

Lejarraja, I., & Walkenhorst, P., (2007). *Diversification by Deepening Linkages with Tourism.* (unpublished).Lewis, A., (1955). *The Theory of Economic Growth.* London: Allen and Unwin.

Lindstrom, B., (2000). Culture and economic development in Aland. In: Baldacchino, G., & Milne, D., (eds.), *Lessons from the Political Economy of Small Islands* (pp. 107–120). The resourcefulness of jurisdiction. London: MacMillan Press Ltd.

Liou, F., & Ding, C., (2002). Small states based on socio economic characteristics. *World Development, 30*(7), 1289–1306.

Looney, R., (1989). Macroeconomic consequences of the size of Third World Countries: With special reference to the Caribbean. *World Development, 17*(1), 69–83.

Lowenthal, D., (2007). Islands, lovers, and others. *Geographical Review, 97*(2), 202–229.

Lucas, R., (1988). On the mechanics of economic development. *Journal of Monetary Economics, 22*(1), 3–42.

Maloney, W., & Montes, R. G., (2001). *Demand for Tourism.* New York: The World Bank.

Marsiglio, S., (2015). Economic growth and environment: Tourism as a trigger for green growth. *Tourism Economics, 21*(1), 183–204.

Marsiglio, S., (2017). On the carrying capacity and the optimal number of visitors in tourism destinations. *Tourism Economics, 23*(3), 632–646.

Marsiglio, S., (2018). On the implications of tourism specialization and structural change in tourism destinations. *Tourism Economics, 24*(8), 945–962.

Mayers, S., & Jackman, M., (2011). Investigating the business cycle properties of tourist flows to Barbados. *The Journal of Public Sector Policy Analysis, 5*, 3–21.

Mazzucato, M., (2018). *The Value of Everything, Making and Taking in the Global Economy.* London, UK: Allen Lane.

McElroy, J., (2006). Small island tourist economies across the life cycle. *Asia Pacific Viewpoint, 47*(1), 61–77.

McElroy, J., & Mahoney, M., (2000). The propensity for dependence in island microstates. *Insula, 9*(1), 32–35.

McElroy, J., & Parry, C., (2010). The characteristics of small island tourist economies. *Tourism and Hospitality Research, 10*(4), 315–328.

McElroy, J., & Sanborn, K., (2005). The propensity for dependence in small Caribbean and Pacific islands. *Bank of Valetta Review, 31*, 1–16.

McGillivray, M., Naudé, W., & Santos-Paulino, A., (2010). Vulnerability, trade, financial flows and state failure in small island developing states. *Journal of Development Studies, 46*(5), 815–827.

McLennan, C., Pham, T., Ruhanen, L., Ritchie, B., & Moyle, B., (2012). Counter-factual scenario planning for long-range sustainable local level tourism transformation. *Journal of Sustainable Tourism, 20*(6), 801–822.

Meade, J. E., et al., (1961). *The Economics and Social Structure of Mauritius—Report to the Government of Mauritius.* London: Methuen.

Metzgen, Q., (1989). *The Demand for International Tourist Services: Theory, Empirics and Policy Perspectives.* Unpublished Ph.D dissertation, Princeton University.

Ministry of General Affairs of Curacao, (2013). *Strategies for sustainable Long Term Economic Development in Curacao.* Retrieved from: https://www.stichtingsmoc.nl/uploads/2013.04.10_Curacao-Report-ook-Isla.pdf (accessed on 4 October 2021).

Modeste, N., (1995). The impact of growth in the tourism sector on economic development: The experience of selected Caribbean countries. *Economia Internazionale, 48*(3), 375–385.

Moore, W., & Whitehall, P., (2005). The tourism area lifecycle and regime switching models. *Annals of Tourism Research, 32*(1), 112–126.

Mowforth, M., & Munt, I., (1998). *Tourism and Sustainability: Development, Globalization and New Tourism in the Third World.* New York, NY: Routledge.

Narayan, K., Narayan, S., Prasad, A., & Prasad, B., (2010). Tourism and economic growth: A panel analysis for Pacific Island Countries. *Tourism Economics, 16*(1), 169–183.

Narayan, P., (2005). Fiji's tourism demand: The ARDL approach to cointegration. *Tourism Economics, 10*(2), 196–206.

Neves-Sequiera, T., & Campos, C., (2005). *International Tourism and Economic Growth: A Panel Data Approach.* Milan: FEEM Working Paper No. 141.05.

Nowak, J., Petit, S., & Sahli, M., (2010). Tourism and globalization: The international division of tourism production. *Journal of Travel Research, 49*(2), 228–245.

Nunkoo, R., Seetanah, B., Jaffur, Z., Moraghen, P., & Sannassee, R. (2020). Tourism and economic growth: a meta-regression analysis. *Journal of Travel Research, 59*(3), 404–423.

O'Rourke, K., (2000). Tariffs and growth in the late 19th Century. *Economic Journal, 110*(463), 456–483.

Ocampo, J., (2002). *Small Economies in the Phase of Globalization.* Lecture delivered at the William G. Demas Memorial Lecture at the Caribbean Development Bank, Cayman Islands.

Oostindie, G., & Klinkers, I., (2001). Knellende Koninkrijksbanden. Het Nederlandse dekolonisatiebeleid in de Caraiben, 1940–2000. Amsterdam: Amsterdam University Press.

Oostindie, G., & Klinkers, I., (2003). *Decolonizing the Caribbean, Dutch Policies in Comparative Perspective.* Amsterdam: Amsterdam University Press.

Oyewole, P., (2001). Prospects for developing country exports of services to the year 2010: Projections and public policy implications. *Journal of Macromarketing, 21*(1), 32–46.

Padilla, A., & Mcelroy, J., (2007). Cuba and Caribbean tourism after Castro. *Annals of Tourism Research, 34*(3), 649–672.

Parra-Lopez, E., & Martinez-Gonzalez, J., (2018). Tourism research on island destinations: A review. *Tourism Review, 73*(2), 133–155.

Pastor, R., & Fletcher, R., (1990). The Caribbean in the 21st century. *Foreign Affairs, 70*, 98.

Patullo, P., (1996). *Last Resorts: The Cost of Tourism in the Caribbean.* London: Cassell.

Payne, A., (2006). The end of green gold? Comparative development options and strategies in the eastern Caribbean banana-producing islands. *Studies in Comparative International Development, 41*(3), 25–46.

Pereira, E., (2018). *Small and smart. An Exploratory Analysis of Economic Institutional Choices of Small Countries and Territories in the Caribbean.* Groningen, the Netherlands: The University of Groningen.

Perez, L., (1974). Aspects of underdevelopment: Tourism in the West Indies. *Science and Society, 37,* 473–480.

Pine, J., & Gilmore, J., (1998). Welcome to the experience economy. *Harvard Business Review.*

Pinker, S., (2018). *Enlightenment Now.* New York: Penguin Books.

Popper, K., (1959). *The Logic of Scientific Discovery.* London: Hutchinson.

Prasad, N., (2003). Small islands quest for economic development. *Asia-Pacific Development Journal, 10*(1), 47–67.

Pratt, S., (2015). The economic impact of tourism in SIDS. *Annals of Tourism Research, 52,* 148–160.

Prebisch, R., (1950). *The Economic Development of Latin America and its Principal Problems.* New York: ECLA, UN Department of Economic Affairs.

Ramkissoon, R., (2002). Explaining differences in economic performance in Caribbean economies. *Paper Presented at the International Conference on Iceland and the World Economy, Small Island Economies in the Era of Globalization.* Center for International Development, Harvard University, Boston, MA. Retrieved from: https://sta.uwi.edu/conferences/financeconference/Conference%20Papers/Session%2030/Explaining%20differences%20in%20Economic%20Performance%20in%20Caribbean%20Economies.pdf (accessed on 4 October 2021).

Read, R., (2001). *Growth, Economic Development and Structural Transition in Small Vulnerable States.* Discussion paper No. 2001/59. New York: UNU/WIDER.

Read, R., (2004). The implications of increasing globalization and regionalism for the economic growth of small island states. *World Development, 32*(2), 365–378.

Reichheld, F., & Sasser, A., (1990). Zero defections: Quality comes to services. *Harvard Business Review, 68,* 105–111.

Richards, G., (1999). Vacations and the quality of life: Patterns and structures. *Journal of Business Research, 44*(3), 189–198.

Richards, G., (2011). Creativity and tourism: The state of the art. *Annals of Tourism Research, 38*(4), 1225–1253.

Ridderstaat, J., (2007). *The Lago Story; the Compelling Story of an Oil Company on the Island of Aruba.* Oranjestad: Editorial Charuba.

Ridderstaat, J., Croes, R., & Nijkamp, P., (2014). Tourism and long-run economic growth in Aruba. *International Journal of Tourism Research, 16,* 472–487.

Ridderstaat, J., Croes, R., & Nijkamp, P., (2016). A two-way causal chain between tourism development and quality of life in a small island destination: An empirical analysis. *Journal of Sustainable Tourism, 24*(10), 1461–1479.

Robinson, E., (1960). The economic consequences of the size of nations. *Proceedings of a Conference Held by the International Economic Association.* London: Macmillan.

Romeu, R., (2008). *Vacation Over: Implications for the Caribbean of Opening US-Cuba Tourism.* International Monetary Fund Working Paper WP/08/162.

Rosenzweig, J., (1988). Elasticities of substitution in Caribbean tourism. *Journal of Development Economics, 29*(1), 89–100.

Roudi, S., Arasli, H., & Akadiri, S., (2019). New insights into an old issue-examining the influence of tourism on economic growth: Evidence from selected small island development states. *Current Issues in Tourism, 22*(11), 1280–1300.

Sachs, J., (2002). The growth performance of small economies. *Paper Prepared for the Conference on Iceland and the World Economy: Small Island Economies in the Era of Globalization.* Center of International Development, Harvard University.

Sachs, J., Warner, A., Aslund, A., & Fischer, S., (1995). Economic reform and the process of global integration. *Brookings Papers Economic Activity, 1*, 1–117.

Sahli, M., & Nowak, J., (2007). Does inbound tourism benefit developing countries? A trade theoretic approach. *Journal of Travel Research, 45*,426–434.

Santos-Paulino, A., (2010). The Dominican republic trade policy review. *World Economy, 33*(11), 1414–1429.

Schahczenski, J., (1990). Development administration in the small developing state: A review. *Public Administration and Development, 10*(1), 69–80.

Scheyvens, R., (2002). Backpacker tourism and third world development. *Annals of Tourism Research, 29*, 144–164.

Scheyvens, R., & Momsen, J., (2008). Tourism in small island states: From vulnerability to strengths. *Journal of Sustainable Tourism, 16*(5), 491–510.

Schianetz, K., Kavanagh, L., & Lockington, D., (2007). The learning tourism destination: The potential of a learning organization approach for improving the sustainability of tourism destinations. *Tourism Management, 28*, 1485–1496.

Schubert, S., Brida, J., & Risso, W., (2011). The impacts of international tourism demand on economic growth of small economies dependent on tourism. *Tourism Management, 32*(2), 377–385.

Schumacher, R., (2012). Adam Smith's theory of absolute advantage and the use of doxography in the history of economics. *Erasmus Journal for Philosophy and Economics, 5*(2), 54–80.

Schwartz, R., (1999). *Pleasure Island: Tourism and Temptation in Cuba.* Lincoln, Nebraska: University of Nebraska Press.

Seetanah, B., (2011). Assessing the dynamic economic impact of tourism for island economies. *Annals of Tourism Research, 38*(1), 291–308.

Seetanah, B., Sannassee, R., & Nunkoo, R. (eds.) (2019). *Mauritius: a Successful Small Island Developing State.* London: Routledge.

Selwyn, P., (1975). *Development Policy in Small Countries.* Beckenham: Croom Helm.

Sequeira, T., & Nunes, P., (2008). Does country risk influence international tourism? A dynamic panel data analysis. *The Economic Record, 84*(265), 223–236.

Sequeria, T., & Campos, C., (2005). *International Tourism and Economic Growth: A Panel Data Approach.* No 2005.141, Working Papers from Fondazione Eni Enrico Mattei.

Setterfield, M., (2002). *The Economics of Demand-Led Growth.* Northampton MA: Edward Elgar.

Seward, S., & Spinrad, B., (1982). *Tourism in the Caribbean, the Economic Impact.* Ottawa, Canada: International Development Research Center.

Shan, J., & Wilson, K., (2001). Causality between trade and tourism: Empirical evidence from China. *Applied Economics Letters, 8*(4), 239–283.

Shand, R., (1980). *The Islands States of the Pacific and Indian Oceans: Anatomy of Development.* Canberra: Australian National University, Development Studies Centre.

Shareef, R., Hoti, S., & McAleer, M., (2008). *The Economics of Small Island Tourism: International Demand and Country Risk Analysis.* Cheltenham, UK: Edward Elgar.

Sharpley, R., (2000). In defense of (mass) tourism. In: Robinson, M., Swarbrooke, J., & Evans, N., et al., (eds.), *Environmental Management and Pathways to Sustainable Tourism* (pp. 269–284). Sunderland: Business Education Publishers Ltd.

Sinclair, T., (1998). Tourism and economic development: A survey. *The Journal of Development Studies, 34*(5), 1–51.

Sinclair, T., & Stabler, M., (1997). *The Economics of Tourism.* London: Routledge.

Smeral, E., (2003). A structural view of tourism growth. *Tourism Economics, 9*(1), 77–94.

Smeral, E., (2012). International tourism demand and the business cycle. *Annals of Tourism Research, 39*(1), 379–400.

Smeral, E., (2016). Tourism forecasting performance considering the instability of demand elasticities. *Journal of Travel Research, 55*(2), 190–204.

Smith, A., (1976). *An Inquiry into the Nature and Causes of the Wealth of Nations.* Chicago: University of Chicago Press.

Song, H., & Wong, K. F., (2003). Tourism demand modeling: A time varying parameter approach. *Journal of Travel Research, 42*(3), 57–64.

Song, H., Witt, S., & Li, G., (2009). *The Advanced Econometrics of Tourism Demand.* Abingdon, UK.: Routledge.

Srinivasan, T., (1986). The costs and benefits of being a small, remote, island, landlocked or mini-state economy. *World Bank Research Observer, 1*(2), 205–218.

Steenge, A., & Van De, S. A., (2010). Tourism multipliers for a small Caribbean Island State: The case of Aruba. *Economic Systems Research, 22*(4), 359–384.

Stiglitz, J., (2012). *The Price of Inequality: How Today's Divided Society Endangers Our Future.* New York: W.W. Norton.

Stiglitz, J., & Greenwald, B., (2014). *Creating a Learning Society: A New Approach to Growth, Development, and Social Progress.* New York: Columbia University Press.

Stiglitz, J., Fitoussi, J., & Durand, M., (2019). *Measuring What Counts: The Global Movement for Well-Being.* New York: The New Press.

Streeten, P., (1993). The special problems of small countries. *World Development, 21*(2), 197–202.

Subramanian, A., & Roy, D., (2003). Who can explain the Mauritian miracle? Maede, Romer, Sachs or Rodrik? In: Rodrik, D., (ed.), *In Search of Prosperity, Analytical Narratives on Economic Growth.* Princeton NJ: Princeton University Press.

Sutton, P., (1987). Political aspect. In: Colin, C., & Tony, P., (eds.), *Politics, Security and development in Small States* (pp. 3–25). London: Allen and Unwin.

Thirlwall, T., (1991). *The Performance and Prospects of the Pacific Island Economies in the World Economy.* Honolulu, HI: University of Hawaii Press.

Tirole, J., (2017). *Economics for the Common Good.* Princeton, NJ: Princeton University Press.

Torres, R., (2002). Cancun's tourism development from a Fordist Spectrum of analysis. *Tourist Studies, 2*(1), 87–116.

Turner, L., (1976). The international division of leisure: Tourism and the third world. *World Development, 4*(3), 253–260.

UNESCO, (1997). https://whc.unesco.org/en/list/ (accessed on 4 October 2021).

UNESCO, (1997). https://whc.unesco.org/en/list/819 (accessed on 4 October 2021).

UNESCO, (1999). https://whc.unesco.org/en/list/ (accessed on 4 October 2021).

Van, S. J., (1978). *Trustee of the Netherlands Antilles*. Zutphen, Netherlands: De Walburg Pers.

Vanegas, M., & Croes, R., (2003). Growth, development and tourism in a small economy: Evidence from Aruba. *International Journal of Tourism Research, 5*(5), 315–330.

Veblen, T., (1994). *The Theory of the Leisure Class*. New York: Dover Publications.

Veenendaal, W., (2013). Democracy in microstates: Why smallness does not produce a democratic political system. *Democratization, 22*(1), 92–112.

Veenendaal, W., (2015). The Dutch Caribbean municipalities in comparative perspective. *Island Studies Journal, 10*(1), 15–30.

Weaver, D., (1988). The evolution of a plantation tourism landscape on the Caribbean Island of Antigua. *Tijdschrift Voor Economische and Sociale Geografie, 79*(5), 319–331.

Weaver, D., (2005). The "plantation" variant of the TALC in the small island Caribbean. In: Butler, R., (ed.), *Tourism area life Cycle: Applications and Modifications* (Vol. 1, pp. 185–197). Clevedon, UK, Channel View Publication.

Webster, A., Fletcher, J., Hardwick, P., & Morakabati, Y., (2007). Tourism and empirical applications of international trade theory: A multi- country analysis. *Tourism Economics, 13*(4), 657–674.

Wignaraja, G., Lezama, M., & Joiner, D., (2004). *Small States in Transition, From Vulnerability to Competitiveness*. London: The Commonwealth Secretariat.

Wilkinson, P. F., (1987). Tourism in small island nations: A fragile dependency. *Leisure Studies, 6*(2), 127–146.

Wilkinson, P., (1989). Strategies for tourism in island microstates. *Annals of Tourism Research, 16*(2), 153–177.

Wilkinson, W., (2016). *What if We Can't Make Government Smaller?* https://www.niskanencenter.org/cant-make-government-smaller/ (accessed on 4 October 2021).

Wint, A., (2002). Competitive disadvantages and advantages of small nations: An analysis of inter-nation economic performance. *Journal of Eastern Caribbean Studies, 27*(3), 1–25.

Winters, A., & Martin, P., (2004). When comparative advantage is not enough: Business costs in small remote economies. *World Trade Review, 3*(3), 347–383.

World Bank, (2005). *A time to Choose, Caribbean Development in the 21st Century*. Washington DC: World Bank.

World Tourism Organization, (2005). *City Tourism & Culture: The European Experience*. A Report of the World Organization and of the Research Group of the European Travel Commission. Madrid: WTO.

World Tourism Organization, (2019). *International Tourism Highlights*. https://www.e-unwto.org/doi/pdf/10.18111/9789284421152 (accessed on 4 October 2021).

WTTC, (2005). *Travel and Tourism Sowing the Seeds of Growth*. The 2005 Travel and Tourism Economic Research. London.

Yang, X., (1994). Endogenous vs. exogenous comparative advantage and economies of specialization vs. economies of scale. *Journal of Economics, 60*(1), 29–54.

Zhang, J., & Jensen, C., (2007). Comparative advantage: Explaining tourism flows. *Annals of Tourism Research, 34*(1), 223–243.

ANNEX 1: TOURISM ARTICLES RELATED TO TOURISM SPECIALIZATION (2000-PRESENT)

Author(s)	Title	Book/Journal
Adamou, A., and Clerides, S. (2009)	Tourism, development, and growth: International evidence and lessons for Cyprus.	Cyprus Economic Policy Review, 3(2), 3–22.
Algieri, B. (2006)	International tourism specialization of small countries.	International Journal of Tourism Research, 8(1), 1–12.
Arezki, R., Cherif, R., and Piotrowski, J. (2009)	Tourism Specialization and Economic Development: Evidence from the UNESCO World Heritage List. International Monetary Fund-Institute and Fiscal Affairs Department.	IMF working paper AP/09/176.
Armstrong, H. W., and Read, R. (2018)	The impact of the 2008 global crisis on small economies in the Caribbean.	Canadian Journal of Latin American and Caribbean Studies, 43(3), 394–416.
Biagi, B., Ladu, M. G., and Royuela, V. (2017)	Human development and tourism specialization. Evidence from a panel of developed and developing countries.	International Journal of Tourism Research, 19(2), 160–178.
Brau, R., Lanza, A., and Pigliaru, F. (2007)	How fast are small tourism countries growing? Evidence from the data for 1980–2003.	Tourism Economics, 13(4), 603–613.
Cerina, F. (2006)	Tourism specialization and sustainability: A long-run policy analysis.	FEEM Working Paper.
Cerina, F. (2007)	Tourism specialization and environmental sustainability in a dynamic economy.	Tourism Economics, 13(4), 553–582.
Chang, C. L., Khamkaew, T., McAleer, M., and Tansuchat, R. (2010)	A panel threshold model of tourism specialization and economic development.	International Journal of Intelligent Technologies and Applied Statistics, 3(2), 159–186.
Chang, C. L., Khamkaew, T., and McAleer, M. (2012)	IV estimation of a panel threshold model of tourism specialization and economic development.	Tourism Economics, 18(1), 5–41.

Author(s)	Title	Book/Journal
Chiang, G. N., Sung, W. Y., and Lei, W. G. (2017)	Regime-switching effect of tourism specialization on economic growth in Asia Pacific countries.	Economies, 5(3), 23.
Chiu, Y.-B., and Yeh, L.-T. (2017)	The threshold effects of the tourism-led growth hypothesis: Evidence from a cross-sectional model.	Journal of Travel Research, 56(5), 625–637.
Croes, R. (2006)	A paradigm shift to a new strategy for small island economies: Embracing demand side economics for value enhancement and long-term economic stability.	Tourism Management, 27(3), 453–465.
Croes, R. (2013)	Tourism specialization and economic output in small islands.	Tourism Review, 68(4), 34–48.
Croes, R., Lee, S. H., and Olson, E. D. (2013).	Authenticity in tourism in small island destinations: a local perspective.	Journal of Tourism and Cultural Change, 11(1–2), 1–20.
Croes, R., Ridderstaat, J., and van Niekerk, M. (2018)	Connecting quality of life, tourism specialization, and economic growth in small island destinations: The case of Malta.	Tourism Management, 65, 212–223.
De Vita, G., and Kyaw, K. S. (2017)	Tourism specialization, absorptive capacity, and economic growth.	Journal of Travel Research, 56(4), 423–435.
Deng, T., Ma, M., and Shao, S. (2014)	Research note: Has international tourism promoted economic growth in China? A panel threshold regression approach.	Tourism Economics, 20(4), 911–917.
Figini, P., and Vici, L. (2010)	Tourism and growth in a cross section of countries.	Tourism Economics, 16(4), 789–805.
Ghalia, T., and Fidrmuc, J. (2018)	The curse of tourism?	Journal of Hospitality and Tourism Research, 42(6), 979–996.
Giannoni, S., and Maupertuis, M. A. (2007)	Is tourism specialization sustainable for a small island economy? A cyclical perspective.	In *Advances in modern tourism research* (pp. 87–105). Heidelberg and New York: Springer, Physica-Verlag.
Giannoni, S., and Maupertuis, M. A. (2007)	Environmental quality and optimal investment in tourism infrastructures: A small island perspective.	Tourism Economics, 13(4), 499–513.
Jackman, M. (2014)	Output volatility and tourism specialization in small island developing states.	Tourism Economics, 20(3), 527–544.

Author(s)	Title	Book/Journal
Jayawardena, C., and Ramajeesingh, D. (2003)	Performance of tourism analysis: A Caribbean perspective.	International Journal of Contemporary Hospitality Management, 15(3), 176–179.
Kim, H. S., and Lee, N. (2010)	Specialization analysis of global and Korean tourism industry: On a basis of revealed comparative advantage.	International Journal of Tourism Sciences, 10(1), 1–12.
Lanza, A., Temple, P., and Urga, G. (2003)	The implications of tourism specialization in the long run: an econometric analysis for 13 OECD economies.	Tourism Management, 24(3), 315–321.
Li, H., Goh, C., and Zhang, Z. (2015)	Is the growth of tourism-specialized economies sustainable? A case study of Sanya and Zhangjiajie in China.	Journal of China Tourism Research, 11(1), 35–52.
Liu, R., Li, L., and Tang, W. (2013)	Is tourism a strategic mainstay industry or independent determinant for economic growth? A review on overseas research on the relationship between tourism and economic growth.	Tourism Tribune/Lvyou Xuekan, 28(5), 35–42.
Logossah, K., and Maupertuis, M. A. (2007)	Does tourism specialization mean sustainable growth for small developing islands?	Journal of Regional and Urban Economics, 1, 35–55.
Marsiglio, S. (2015)	Economic growth and environment: Tourism as a trigger for green growth.	Tourism Economics, 21(1), 183–204.
Marsiglio, S. (2017)	On the carrying capacity and the optimal number of visitors in tourism destinations.	Tourism Economics, 23(3), 632–646.
Marsiglio, S. (2018)	On the implications of tourism specialization and structural change in tourism destinations.	Tourism Economics, 24(8), 945–962.
McElroy, J. L., and Hamma, P. E. (2010)	SITEs revisited: Socioeconomic and demographic contours of small island tourist economies.	Asia Pacific Viewpoint, 51(1), 36–46.
Nowak, J. J., Petit, S., and Sahli, M. (2010)	Tourism and globalization: the international division of tourism production.	Journal of Travel Research, 49(2), 228–245.
Oberst, A., and McElroy, J. L. (2007)	Contrasting socio-economic and demographic profiles of two, small island, economic species: MIRAB versus PROFIT/SITE.	Island Studies Journal, 2(2), 163–176.
Pablo-Romero, M. D. P., and Molina, J. A. (2013)	Tourism and economic growth: A review of empirical literature.	Tourism Management Perspectives, 8, 28–41.

Author(s)	Title	Book/Journal
Pena-Boquete, Y., and Pérez-Dacal, D. (2012)	Effects of Tourism Wages and employment for the Spanish regions: Seasonality versus Tourism Specialization.	International Conference on Regional Science.
Perez-Dacal, D., and Pena-Boquete, Y. (2013, November)	A regional analysis of Tourism Specialization in Spain. In *ERSA conference papers* (No. ersa13p1238).	European Regional Science Association.
Pérez-Dacal, D., Pena-Boquete, Y., and Fernández, M. (2014)	A measuring tourism specialization: A composite indicator for the Spanish regions.	Almatourism-Journal of Tourism, Culture, and Territorial Development, 5(9), 35–73.
Po, W. C., and Huang, B. N. (2008)	Tourism development and economic growth-a nonlinear approach.	Physica A: Statistical Mechanics and its Applications, 387(22), 5535–5542.
Resende-Santos, J. (2019)	Cape Verde and the risks of tourism specialization: The tourism option for Africa's small states.	Journal of Contemporary African Studies, 37(1), 148–168.
Rey-Maquieira, J., Lozano, J., and Gomez, C. M. (2009)	Quality standards versus taxation in a dynamic environmental model of a tourism economy.	Environmental Modeling and Software, 24(12), 1483–1490.
Ridolfi, E., Pujol, D. S., Ippolito, A., Saradakou, E., and Salvati, L. (2017)	An Urban Political Ecology approach to local development in fast-growing, tourism-specialized coastal cities.	Tourismos, 12(1), 171–210.
Risso, W. A. (2018)	Tourism Specialization, Income Distribution, and Human Capital in South America.	In *Tourism-Perspectives and Practices*. IntechOpen.
Risso, W. A. (2018).	Tourism specialization, income distribution, and human capital in South America.	In *Tourism*. IntechOpen.
Romão, J., and Neuts, B. (2017).	Territorial capital, smart tourism specialization and sustainable regional development: Experiences from Europe.	Habitat International, 68, 64–74.
Romão, J., and Nijkamp, P. (2019)	Impacts of innovation, productivity, and specialization on tourism competitiveness-a spatial econometric analysis on European regions.	Current Issues in Tourism, 22(10), 1150–1169.
Rontos, K., Syrmali, M. E., Vavouras, I., Karagkouni, E., and Salvati, L. (2017)	Tourism in time of crisis: Specialization, spatial diversification, and potential to growth across European regions.	Tourismos, 12(2), 180–206.

Author(s)	Title	Book/Journal
Sequeira, T. N., and Campos, C. (2007)	International tourism and economic growth: A panel data approach.	In *Advances in Modern Tourism Research* (pp. 153–163). Heidelberg and New York: Springer, Physica-Verlag.
Šergo, Z., and Gržinić, J. (2019)	Kaldor's income distribution and tourism specialization: evidence from selected countries.	In *Approaches in Tourism Modeling* (pp. 165–196). Sciendo.
Serra, D. (2000)	The identity and infra-regional specialization of tourism in small isolated economies.	Téoros, Revue de Recherche en Tourisme, 19(3), 20–27.
Singh, D. R. (2008)	Small island developing states (SIDS): Tourism and economic development.	Tourism Analysis, 13(5–6), 629–636.
Soulie, J., and Valle, E. (2014)	Trade effects of specialization in tourism: An inter-regional input-output model of the Balearic Islands.	Tourism Economics, 20(5), 961–985.

ANNEX 2: LEAKAGE CALCULATION

The formula for calculating the leakage is as follows:

$$\text{Leakage}_t = \frac{\left(\text{Tourism receipts}_t - \text{Tourism imports}_t\right)}{\text{Nominal GDP}_t} \times 100\% \qquad (1)$$

Tourism receipts and nominal GDP data are from the Central Bank of Aruba (www.cbaruba.org). For the estimation of tourism-related imports, I used information on consumption weights provided by the Central Bureau of Statistics in Aruba (see Table 5.7). The overall weight of residents is equal to the sum of all sectors (= 10,000). Tourists are assumed here to consume only in a number of sectors, and the sum of these sectors is estimated at 4,728.2. Relating the tourists' consumption to that of residents of Aruba leads to a ratio of 32.10% or about a third of the overall consumption is attributed to tourism.[1] The latter weight is then used to estimate the value of imports attributed to tourism (see Table 5.6). On average, the leakage was estimated at 43.8% between 2010 and 2017.

TABLE 5.6 Consumption Weights of Residents and Tourists

Consumption Sectors	Overall	Resident	Tourists
Food and non-alcoholic beverages	1087.7	1087.7	1087.7
Alcoholic beverages and tobacco	78.2	78.2	78.2
Clothing and food wear	276.8	276.8	–
Housing	2,522.20	2,522.20	–
Household operation	929.2	929.2	–
Health	223.1	223.1	–
Transport	1,257.40	1,257.40	1,257.40
Communications	849.6	849.6	849.60
Recreation and culture	1013.3	1013.3	1,013.30
Education	98	98	–
Restaurants and hotels	442	442	442
Miscellaneous goods and services	1,222.6	1,222.6	–
Total	**10,000.0**	**10,000.0**	**4,728.2**
Ratio Tourist Versus Residents (in %):			**32.10%**

Source: Central Bureau of Aruba and author's calculations.

---

[1] Ratio tourist versus residents $= \dfrac{4{,}728.2}{10{,}000 + 4{,}728.2} \times 100\% = 32.10\%$

TABLE 5.7   Estimation of Tourism Leakage Percentage

		2010	2011	2012	2013	2014	2015	2016	2017	Average
A	Tourism receipts	2,226.70	2,409.60	2,501.40	2,666.00	2,861.30	2,946.60	2,891.70	3,078.40	
B	Imports (non-oil)	1,470.20	1,786.20	1,760.00	1,860.60	1,758.20	1,672.80	1,593.50	1,705.10	
B1(= B x 0.3210)	Tourism imports	471.98	573.42	565.33	597.31	564.43	537.02	511.56	547.39	
D(=A − B1)	Net	1,754.72	1,836.18	1,936.07	2,068.69	2,296.87	2,409.58	2,380.14	2.531.01	
E	Nominal GDP	4,279	4,564	4,537	4,836	4,950	5,226	5,309	5,471	
(B+C)/E	Leakage percentage	41.0%	40.2%	42.7%	42.8%	46.4%	46.1%	44.8%	46.3%	43.8%

Source: Central Bank of Aruba and authors calculations.

INDEX

For Product Safety Concerns and Information please contact our EU
representative GPSR@taylorandfrancis.com
Taylor & Francis Verlag GmbH, Kaufingerstraße 24, 80331 München, Germany

www.ingramcontent.com/pod-product-compliance
Lightning Source LLC
Chambersburg PA
CBHW060238220326
41598CB00027B/3975